Research Reports ESPRIT

Subseries
Project 322 · CAD Interfaces (CAD*I)
Volume 7

Subseries Editors:
I. Bey, Kernforschungszentrum Karlsruhe GmbH
J. Leuridan, Leuven Measurement and Systems

Edited in cooperation with
the Commission of the European Communities

H. Grabowski R. Anderl M. J. Pratt (Eds.)

Advanced Modelling for CAD/CAM Systems

Springer-Verlag
Berlin Heidelberg New York London Paris
Tokyo Hong Kong Barcelona Budapest

Volume Editors

Hans Grabowski
Reiner Anderl
Institut für Rechneranwendung in Planung und Konstruktion
Universität Karlsruhe
Postfach 69 80, W-7500 Karlsruhe, FRG

Michael J. Pratt
Dept. of Applied Computing and Mathematics
Cranfield Institute of Technology (CIT)
Cranfield, Bedford MK43 OAL, UK

ESPRIT Project 322: CAD Interfaces (CAD*I) belongs to the Research and Development area "Computer-Aided Design and Engineering (CAD/CAE)" within the Subprogramme 5 "Computer-Integrated Manufacturing (CIM)" of the ESPRIT Programme (European Strategic Programme for Research and Development in Information Technology) supported by the European Communities.

ESPRIT Project 322 has been established to define the most important interfaces in CAD/CAM systems for data exchange, data base, finite element analysis, experimental analysis, and advanced modelling. The definitions of these interfaces are being elaborated in harmony with international standardization efforts in this field.

Partners in the project are:
Bayerische Motorenwerke AG / FRG · CISIGRAPH / France · Cranfield Institute of Technology / UK · Danmarks Tekniske Højskole / Denmark · Estudios y Realizaciones en Diseño Informatizado SA (ERDISA) / Spain · Gesellschaft für Strukturanalyse (GfS) mbH / FRG · Katholieke Universiteit Leuven / Belgium · Kernforschungszentrum Karlsruhe GmbH / FRG · Leuven Measurement and Systems / Belgium · NEH Consulting Engineers ApS / Denmark · Rutherford Appleton Laboratory / UK · Universität Karlsruhe / FRG.

CR Subject Classification (1991): I.3.5, J.6

ISBN 3-540-53943-3 Springer-Verlag Berlin Heidelberg New York
ISBN 0-387-53943-3 Springer-Verlag New York Berlin Heidelberg

Publication No. EUR 13536 EN of the
Commission of the European Communities, Scientific and Technical Communication Unit,
Directorate-General Telecommunications, Information Industries and Innovation,
Luxembourg

Printed in Germany

Printing and Binding: Weihert-Druck GmbH, Darmstadt
2145/3140 – 543210 – Printed on acid-free paper

Table of Contents

CAD*I Project Overview

During the past 25 years computers have been introduced in industry to perform technical tasks such as drafting, design, process planning, data acquisition, process control and quality assurance. Computer-based solutions, however, are still in most cases single isolated devices within a manufacturing plant.

Computer technology is evolving rapidly, and the life cycle of todays' products and production methods is shortening, with continuously increasing requirements from customers, and a trend to market interrelations between companies at a national and international level. This forces a growing need for efficient storage, retrieval and exchange of information. Integration of information is urgent within companies to interconnect departments which used to work more or less on their own. On the other hand direct communication with outside customers, suppliers and partner institutions will often determine the position of an enterprise among its competitors. In this sense, Computer Integrated Manufacturing (CIM) is the key of today for the competitiveness of tomorrow. But the realization of a future-oriented CIM concept is not possible without powerful, widely accepted and standardized interfaces. They are the vital issue on the way to CIM. They will contribute to harmonizing data structures and information flows and will play a major role in open CIM systems. Standardized interfaces allow for:

- Access to data produced and archived on computing equipment which is no longer in active use;
- Communication between hardware and software from different vendors;
- Paperless exchange of information.

ESPRIT Project 322 "CAD Interfaces" (CAD*I) started in 1984 is a five-year research and development programme on CAD interfaces with the aim of defining some missing interface specifications in the environment of computer aided design (CAD) systems for mechanical engineering. Parts design and CAD are the starting point in the design and manufacturing process, and can also be considered as a starting point for information generation and data exchange.

Based on the results and using the experiences of former national standardization initiatives like IGES, VDAFS or SET, the CAD*I project team aimed from early in the project to contribute to the first international standard for product data exchange, because only an internationally accepted standard interface will fulfill the requirements of European industry.

The standardization work in CAD data exchange at international level is performed through ISO/TC184/SC4 under the name STEP: Standard for the Exchange of Product Model Data. CAD*I has had a large influence on the STEP definitions especially for the exchange of geometry and shape information (curves, surfaces and solid models), the interface to Finite Element Analysis programmes and drafting information.

This report is one of a series of similar books which summarize the wealth of results achieved during the five years of ESPRIT Project CAD*I.

CAD Interfaces

The main results are:

- Vendor independent interface consisting of a neutral file specification and corresponding pre- and post-processors for many commercial CAD systems have been defined, developed and tested. The CAD*I specifications for geometry and shape representation (curves, surfaces and solids) are clearly visible in the first international draft proposal standard. The processors are in practical use in several European and national projects. European system vendors have begun to integrate these results into their products.
- A general standard specification of a neutral file for exchanging finite element data has been developed and implemented. Tests have been performed with the interface processors for several FEM packages available on the market. In addition CAD models were transferred to finite element systems using the CAD*I neutral file. The results of this work have already appeared on the European market.
- New and improved data acquisition and analytical procedures for dynamic structural analysis have been specified and tested on complex real structures. Also, powerful tools for the intelligent integration (link) of

experimental and analytical results in structural design have been developed, tested and merged into software products now available on the market. These results are visible in recent commercial products.

- Some new methods have been developed to enhance the communication interface in CAD/CAE systems. Future users of this kind of system will be able to enter information to the systems by handsketching input or by technical terms from using design language instead of formal geometrical descriptions. First implementations have been successful; they are based on levels of internal interfaces using a product model.
- A neutral database interface based on the CAD*I neutral file format has been developed to handle archiving and retrieval of product information in a database. A set of standard access routines has been written and tested with existing CAD systems and a widely used commercial relational database management system. The introduction of these results into marketable products is on the way.
- An information model for the description of technical drawings has been developed: the CAD*I drafting model. This information model represents the highest level of sophistication within the level concept of the drafting model of the STEP specification.

A total of about 150 person-years of research and development effort has been spent on the project. The CAD*I project involved 12 partners in 6 countries of the European Community.

As project manager since 1985 I would like to express my appreciation to the co-manager J. Leuridan and the fifty or more people working in and on the project for their engagement to reach the originally stated goals. In addition I would like to pay special tribute to Mrs. P. MacConaill and R. Zimmermann from the Commission of the European Communities and to the reviewers G. Enderle (†), E.A. Warman and H. Nowacki for their cooperative support. Special acknowledgement is due also to Mrs. U. Frey for running the administrative part of the project and for her contributions to forming the spirit of the CAD*I team.

I. Bey, CAD*I Project Manager

1. Introduction

Reiner Anderl

The Advanced Modelling part of the CAD*I project aimed at the development of a new generation of modelling techniques as a basic functionality of future CAD/CAM systems. The methodology and concepts for advanced modelling techniques, their availability in the communication interface of a CAD/CAM system and their influence on internal interfaces in the software architecture of a CAD/CAM system are fundamental results of advanced modelling work. These results form the basis for the development of a new generation of CAD/CAM systems which are called product modelling systems.

CAD/CAM systems today mainly support the geometric description of a technical part or its description as a technical drawing. Advanced geometric modelling capabilities deal with parametric design functions embedded into CAD/CAM systems. However, development strategies for future CAD/CAM systems are directed toward the following:

1. The development of product modelling systems

   and

2. the development of integrated systems based on CAD, CAP (Computer Aided Planning), CAM and other CIM (Computer Integrated Manufacturing) functionalities.

The basic idea of these development strategies is to represent all the product characteristics in a computer internal representation and to provide the modelling techniques which allow the definition and manipulation of those product characteristics. The computer internal representation of the product characteristics is also being called product model and covers all the product data which occur during the whole product life cycle. The advanced modelling techniques process this product model data based on the input coming from graphical-interactive man-machine communication. This communication process for product modelling however provides communication techniques which are derived from the language and the methods used by designers.

The advanced modelling part of the CAD*I project provides a methodology for product modelling techniques and a multi-level concept represented by a matrix where modelling levels are listed in rows and design techniques are listed in columns. Based on this multi-level concept functions for technical modelling have been developed and the internal interface specification as part of the product modelling system architecture has been worked out. It should also be mentioned that results of the advanced modelling work have been brought into the development work for the Standard for the Exchange of Product Model Data (STEP) within the ISO (International Standardization Organization) TC 184 SC4 .

2. Methodology of Design

Stefan Rude

The work of a designer can be conceived as discovering a solution for a given problem. Moreover, the result of the design is a technical product that functions and can be manufactured.

This work is decisively influenced by the environment in which the designed object will be used. In addition, available materials, production methods, etc. must also be considered.

Intelligence is required to perform this work. This can be understood as both technical knowledge and the capability to apply this knowledge to a given problem. Often the knowledge exists in the form of sample patterns that present mental pictures of a design. Finding a solution consists of selecting a pattern. This is an important task because the selected pattern rarely matches the problem exactly: it must be adapted to some extend to the particular problem. The skill in finding a solution resides in the choice of a pattern which requires the least amount of adaptation.

2.1 General Problem Solving Principles

The design process consists of numerous activities which can be described using systems analysis methods /GRR-88a/.

The design process then appears as a process of stepwise refinement in which the following major activities occur over and over again:

- finding a mental solution
- representing the solution
- analysing the solution.

Figure 2.1-1 shows a possible method for solving design problems. Here, the additional activity of "confrontation with the problem" preceeds the above three major activities. These are furthermore followed by a "final decision making" step.

Design problems are solved within that scheme by recursive and iterative steps. Starting with a given problem, the problem solving cycle will be traversed several times, starting with the subdivision of the total problem into a set of subproblems. At the same time, the designer attempts to obtain more concrete solutions from more abstract descriptions of the problem, i.e. the represented solutions will be different sets of information on different levels of abstraction. Methods to find a solution are conventionally distinguished into intuitive and into discursive ones. The analysis of a solution involves calculation, simulation and test methods.

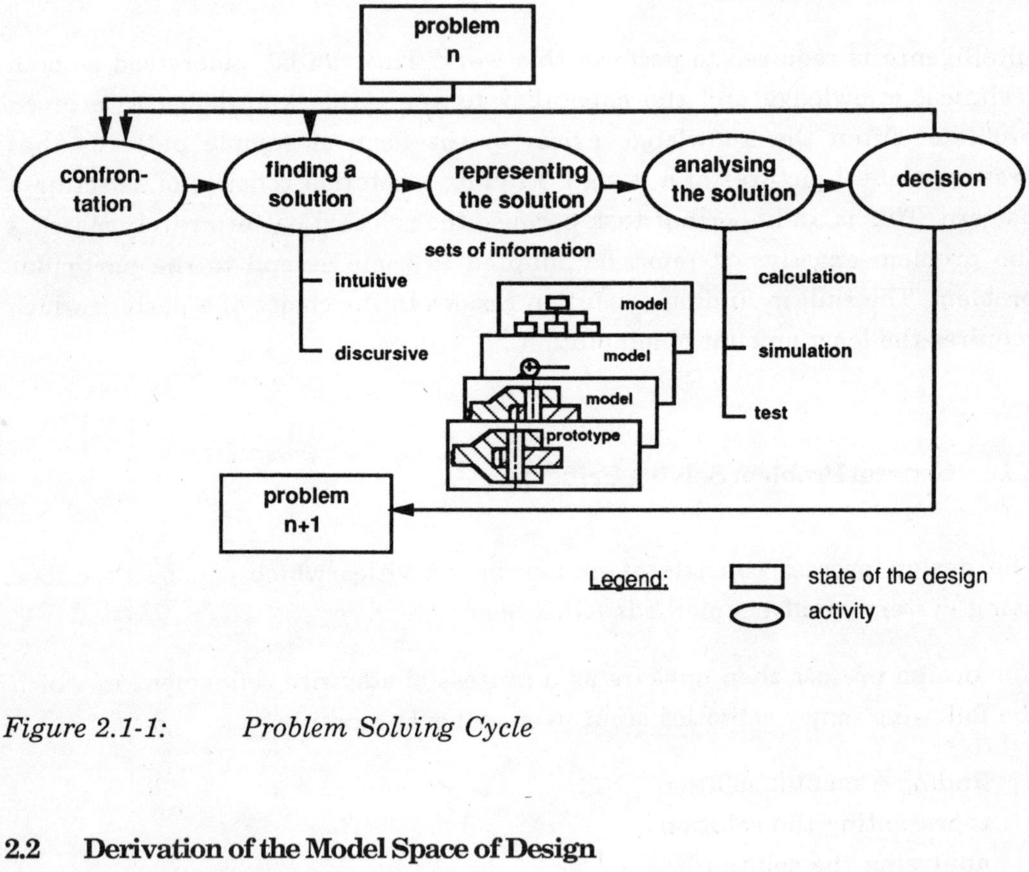

Figure 2.1-1: *Problem Solving Cycle*

2.2 Derivation of the Model Space of Design

A refinement of a solution will be done step by step from a very abstract representation to more concrete ones (top-down-design, see Figure 2.2-1). The scheme described above can be used in each design phase.

The results obtained at the end of each design phase are the basis for the following phase. The result of the function definition phase is an abstract description of a product. This abstract product description is a function structure which describes how an overall function of the product can be realised in terms of a combination of subfunctions.

The choice of physical principles for realising the design involves finding a solution principle for each element of the function structure.

In the preliminary design phase these physical principles are used for the definition of initial shape characteristics. The result is a complete design scheme.

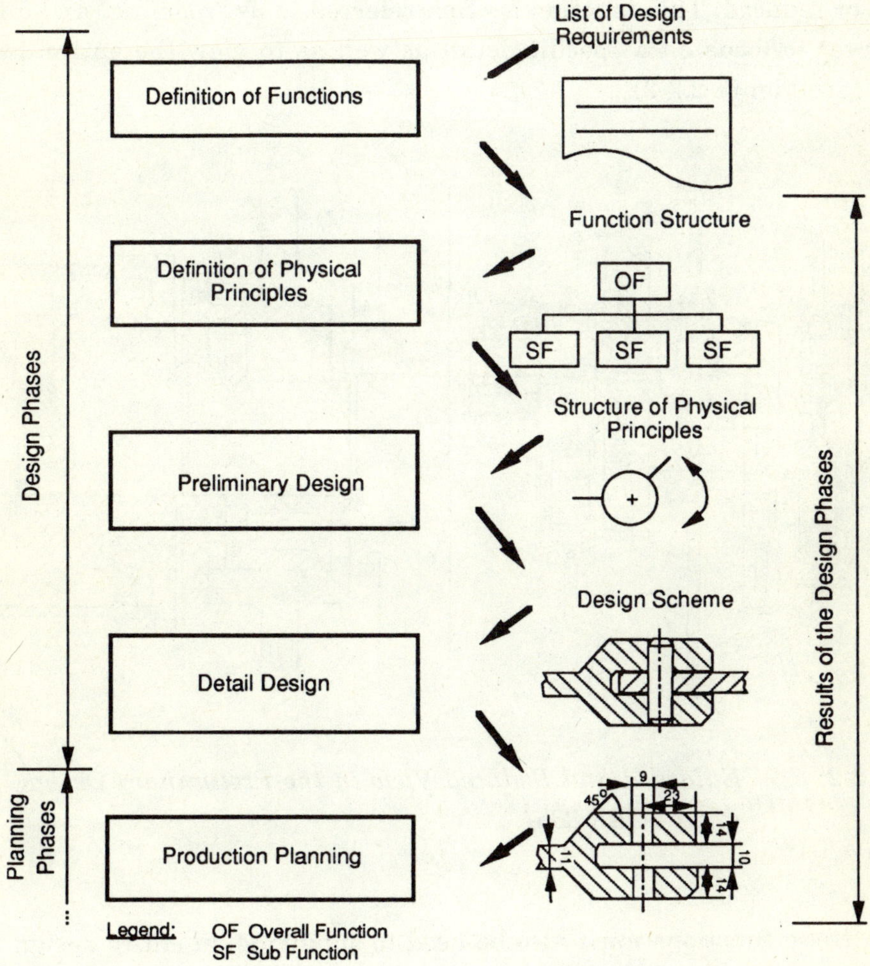

Figure 2.2-1: *Design Phases and their Results*

During the detailed design phase final product descriptions are worked out. These documents are the basis for NC-programming, part list generation, process planning, etc.

Returning to an already concluded phase may become necessary as additional information from later phases becomes available. A revision of a decision reached in an earlier phase might also be necessary. Therefore, the design process also carries some aspects of a bottom-up approach. Real design tasks however always switch between a top-down and a bottom-up strategy depending on the results expected on the current level of abstraction.

In addition to this aspect of design, the drawings of a given structure can be enlarged or reduced. This function is often referred to as "zooming" and allows the observer to focus on a specific detail as well as to view the entire design /RUT-85/ (see Figure 2.2-2).

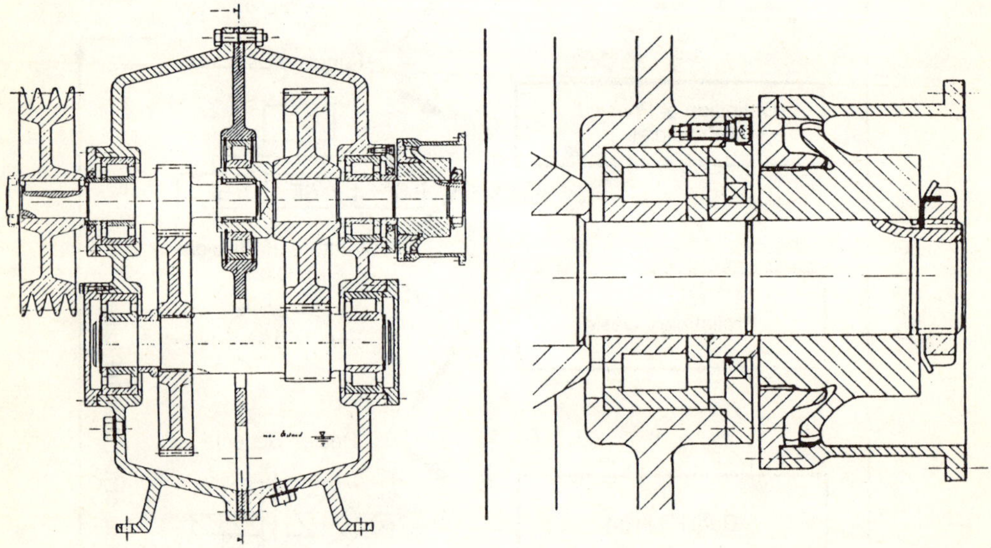

Figure 2.2-2: Enlarged and Reduced View in the Preliminary Design Phase /GRR-88b/

However, those functions must also be used to subdivide an entire design into several subproblems, which can be solved by different designers. It is therefore essential that focusing onto a specific detail of a design also locks its environment.

Additionally, focusing on details does not occur only within the preliminary design phase, but also occurs within the more abstract design phases. An example of the focusing process during the selection of physical principles is shown in Figure 2.2-3.

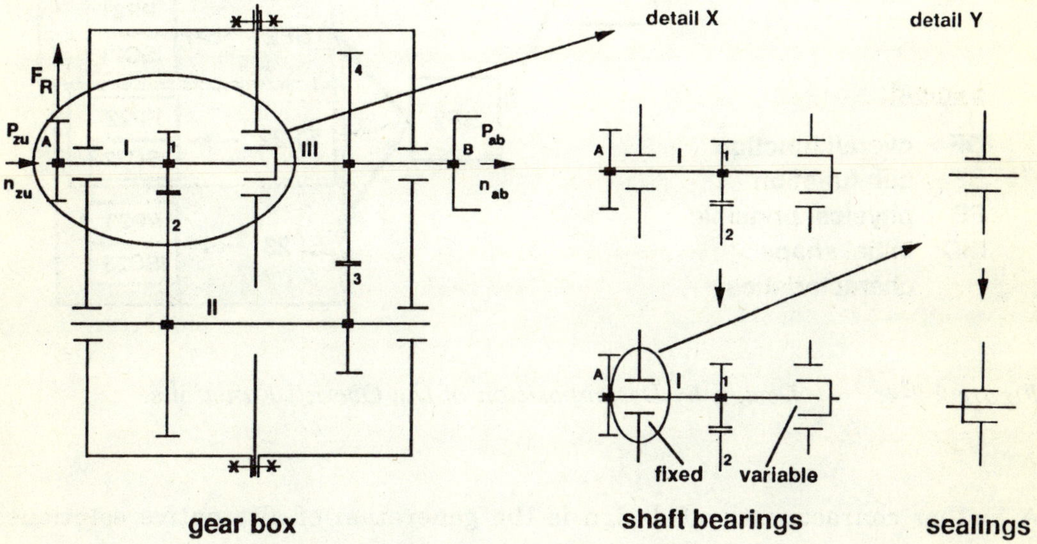

Figure 2.2-3: *Enlarged and Reduced View During the Phase of Selecting Physical Principles /GRR-89/*

In the functional design phase /GRB-89/ views onto the problem can be similarly structured into more detailed and global views, see Figure 2.2–4.

Here a mechanism is proposed, which yields a function structure in the form of a tree which reflects the function-subfunction hierarchy, the solution principles and the corresponding initial shape characteristics.

However, this tree will contain duplicate subfunctions, solution principles and initial shape characteristics. These must be integrated and combined up the tree so as to synthesize the overall function while taking into account physical and geometric compatibility constraints.

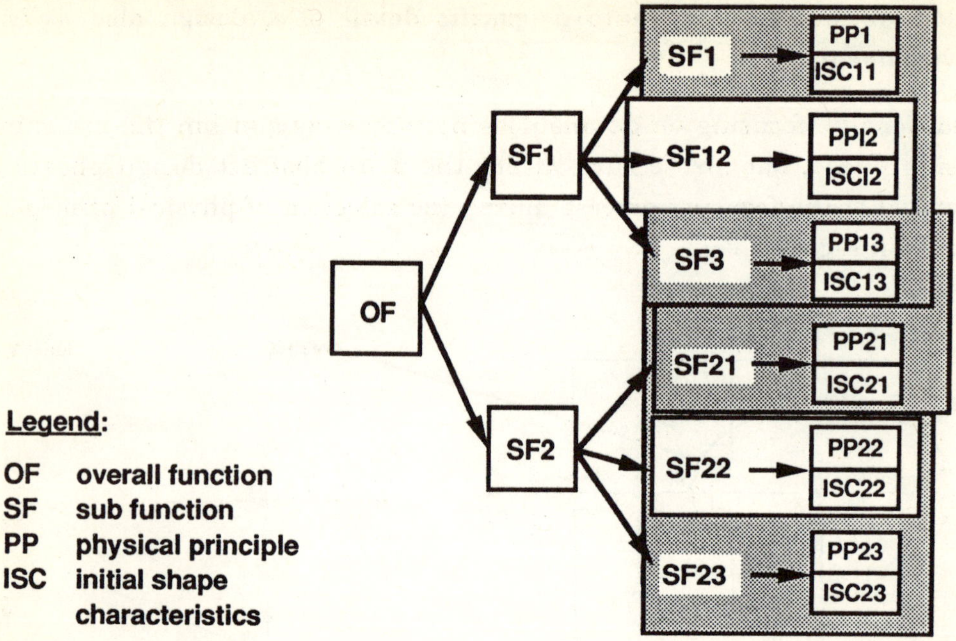

Figure 2.2-4: Design by Decomposition of the Overall Functions

A further characteristic of design is the generation of alternative solutions based on given solutions. That means that unless a problem is solved in principle mechanisms occur which are able to vary a solution in a way that design alternatives are created. An example is the classification scheme for shaft-hub joints of Pahl/Beitz /PAB-86/. A principal solution is given, but alternatives can be found by varying the parameters of the solution (see Figure 2.2-5).

These design characteristics can be taken to build a design model space whose dimensions are the degree of concreteness, the degree of detail and the solution alternatives /GRR-88b/. Figure 2.2-6 illustrates this design model space. Additionally, some typical tasks of design work are represented in the figure as transitions from one state of design to another.

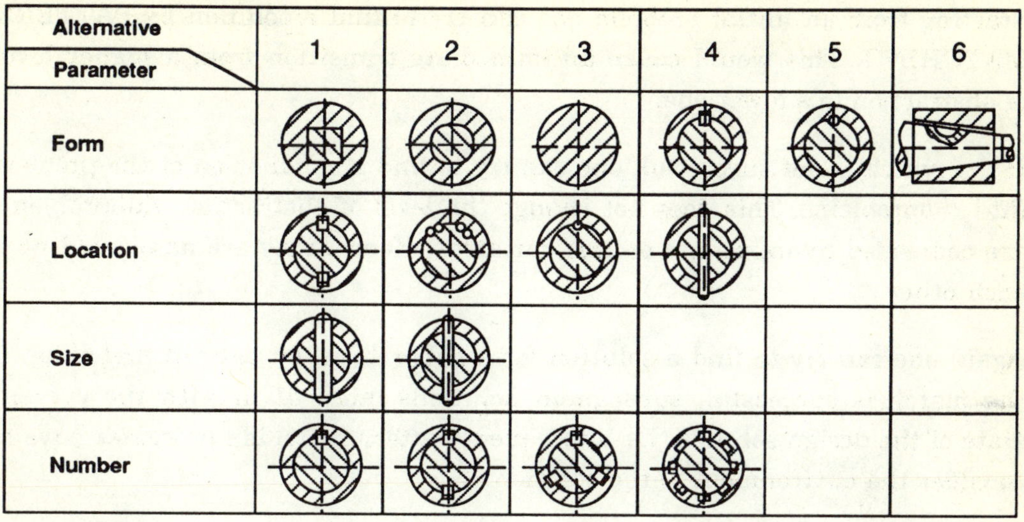

Figure 2.2-5: *Design Alternatives /PAB-86/*

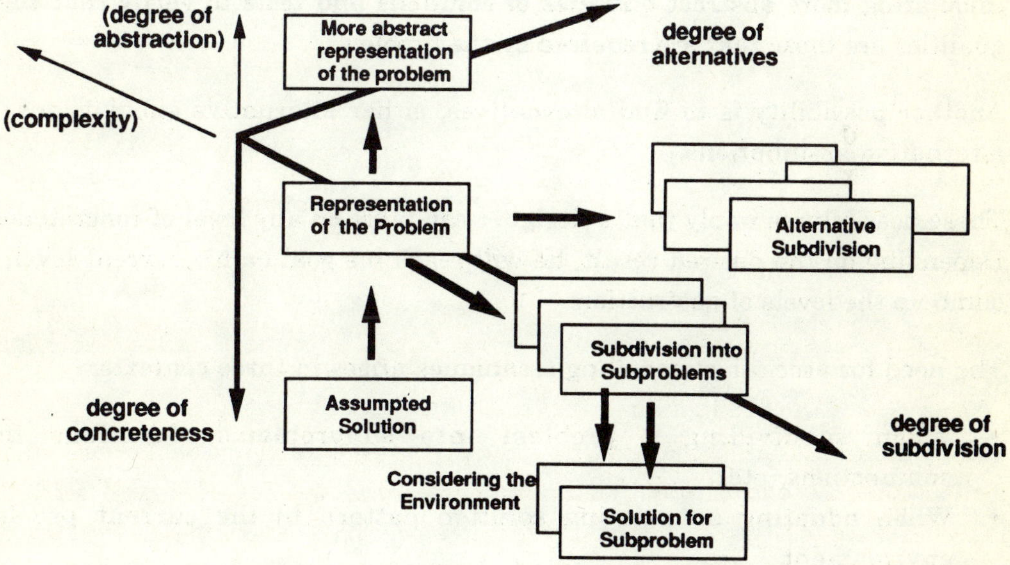

Figure 2.2-6: *Model Space of Design*

Starting from an initial problem one can try to find a solution by PATTERN MATCHING. This would cause an immediate transition from a higher level of abstraction to a lower one.

If the match is not successful, one can try to find a subdivision of the problem into subproblems. This does not change the level of abstraction. Subproblems are connected by input and output variables. Hence, they are associated with each other.

Again one can try to find a solution for a subproblem by pattern matching. If the match is successful, subsequent solutions must fit in with the current state of the design solution, i.e. subsequent pattern matching processes have to consider the environment of the solution.

If the solutions are found without regarding the environment, the compatibility of solutions must be tested later on in the design process.

Another principle for finding solutions is to make assumptions. This requires calculating more abstract qualities of solutions and tests to verify that these qualities are those that are required by the problem.

Another possibility is to find alternatives, either alternative subdivisions, or alternative assumptions.

These possibilities imply that a designer can start on any level of concreteness. Depending on the desired result, he will reach his goal on his current level, or climb up the levels of abstraction.

The need for asociative modelling techniques arises in three contexts:

- When subdividing a problem into subproblems, functions into subfunctions, etc.
- When adapting an existing solution pattern to the current problem environment.
- When verifying that a partial solution is compatible with the current state of the design.

3. Overview of the Concept for Advanced Modelling

Stefan Rude

3.1 The Integrated Product Model

There are two ways of representing the design phases as a computer internal model (Figure 3.1-1).

A first step involves the definition of several different model, called phase models: each model contains all the information needed for the corresponding phase of the design process.

The advantages of this approach are:

- Simplification of the different models,
- reduction of the amount of data, and
- good response times for interactive use.

To realize this approach functions must be provided to transfer information between different phase models. However, this transformation between the models cannot be done without loss of information.

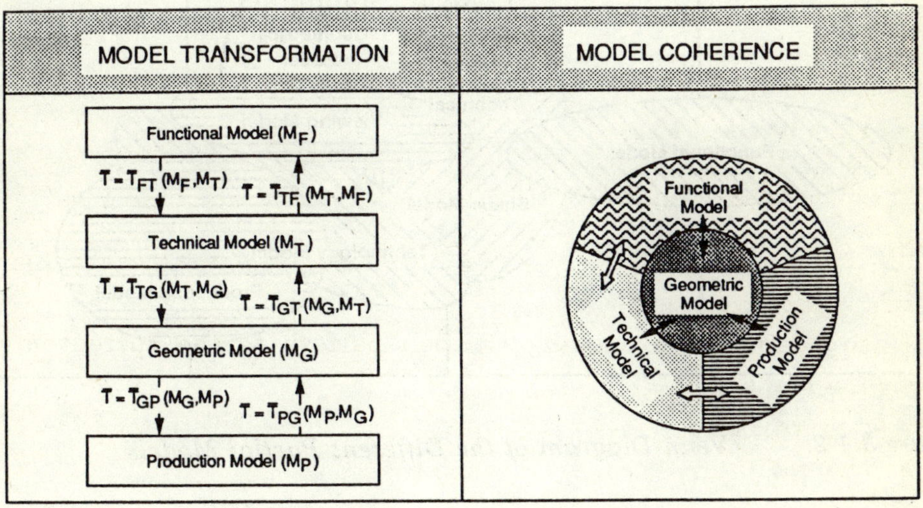

Figure 3.1-1: Comparison between Model Transformation and an Integrated Model /GRS-84/

This disadvantage can be avoided by the second approach /SEI-85/. This approach is based on an integrated product model, which contains all the information for each phase of design and manufacture.

Such an integrated product model covers all technical product information and includes semantic relations. This is necessary for the provision of an integrated support of all technical phases of the design process.

The integrated product model is structured into several logical clusters, i.e. partial models. A partial model is understood as a set of entities which logically belong together.

Distinct partial model have disjoint entity sets but overlapping sets of relationships.

The integrated product model can be represented as a Venn diagram of its partial models (Figure 3.1-2).

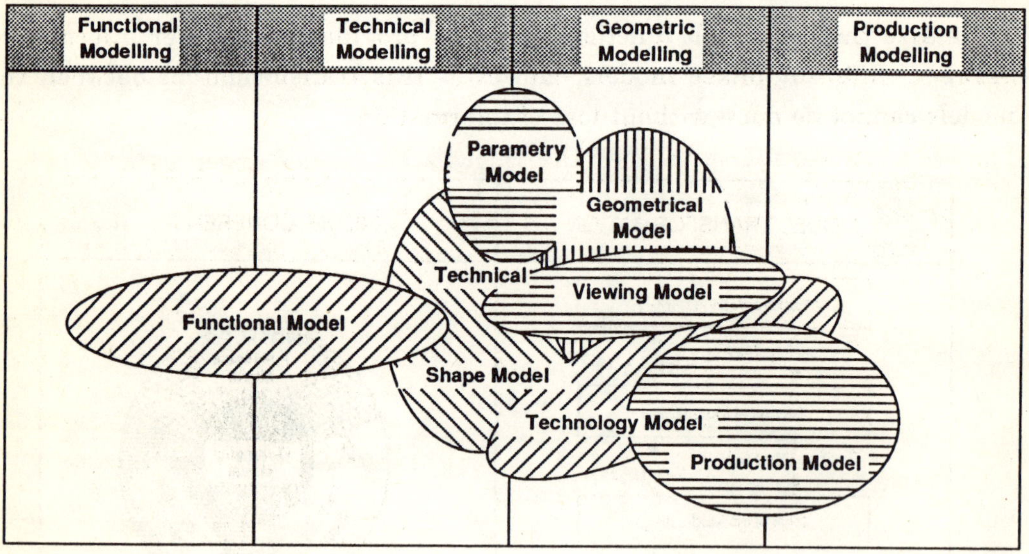

Figure 3.1-2: Venn Diagram of the Different Partial Models

The partial models correspond to different design phases.

The partial models are /SEI-85, AND-85, KAN-88, BEN-90, HEI-90/:

The **Functional Model** involves modelling the overall function and its decomposition into sub functions as well as the corresponding physical principles.

The **Technical Shape Model** represents the shape and structure information, which cannot be described by analytical geometry. It is the technical topology of the product. The kinematics of technical objects are also described in the technical shape model.

The **Geometric Model** describes the product geometry and topology according to the boundary representation method.

The **Technological Model** comprises the product's technological data like material, tolerances, surface texture, etc. This will be referred to as micro geometry in the following text.

The **Parametric Model** contains information about mathematical relationships between product variables. It is need to store parametric parts.

The **Viewing Model** contains all the information needed to generate views and engineering drawings.

The **Production Model** comprises production related data like the available machines and tools, intermediate states of parts during the manufacturing process, etc.

3.2 Modelling Levels

Starting from the integrated product model modelling levels were developed which allow a unified product development (Figure 3.2-1).

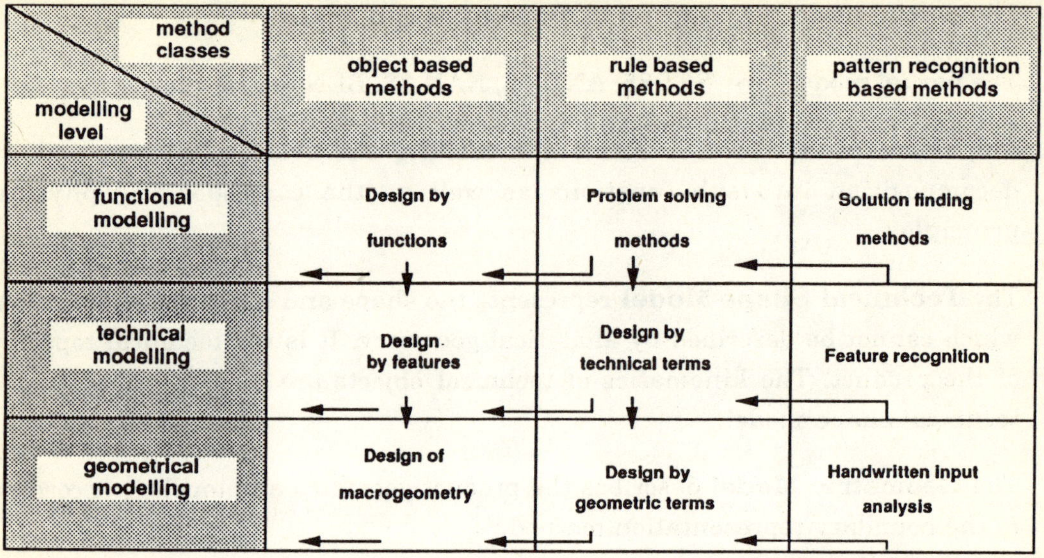

Figure 3.2-1: *Matrix of Modelling Levels and Classes of Methods*

The modelling levels are:

- The geometric modelling level. The task is to achieve the geometric representation of parts. It is important to distinguish between geometry, topology, and geometric constraints such as angularity or parallelity.
- The technical modelling level. The purpose at this level is to give the geometry an engineering meaning based on the designer's input.
- The functional modelling level. Its role is to link an object's preconditions, functions, physical equations, and physical structures to the technical elements created at the technical modelling level.

With respect to specification of method classes, the following distinctions can be made:

- Object based modelling methods
  Based on the user input, this modelling method uses class representations of solutions (either as programs or data structures or both) and is able to insert these into the current data structure.

Considering conventional modelling methods at the **geometric level**, the modelling of macro geometry can be seen as the complete specification of a geometric element (e.g. insert cylinder with length = 100 and diameter = 20).

At the **technical level**, several methods can be used to perform modelling operations. These involve the complete specification of technical elements (e.g. insert bore hole with diameter = 5 and depth = 10).

In **functional modelling**, technical solutions are obtained in terms of functions. This is achieved through combinations of technical elements (e.g. insert a link between the shaft and flange: spline length = 10, number of splines = 20).

- Rule based modelling methods

  Here element attributes are determined from the user input.

  At the **geometric modelling** level, associative geometric techniques enable position, orientation, and dimensions of geometric elements to be determined (for example, insert planar face parallel to "some user selectable entity").

  At the **technical level**, modelling techniques can be provided which determine the attributes of technical elements as well as the position, orientation and dimensions of their geometric representations (e.g. insert bearing adjacent to <ident.>).

  At the **functional level**, methods for the specification of functional attributes must be provided (for example, the functional output of one function is equal to the functional input of a second function).

- Modelling methods based on pattern recognition generate information which will be used at the next higher level.

  The **geometric level** consists of all methods for the analysis of geometric inputs. Consider the example of the input system for freehand sketches which has to perform two tasks: the analysis of the type of geometric element and that of the relations between the recognized geometric elements.

  The **technical level**: this involves methods for the recognition of technical semantics of geometry such as the recognition of form features for the purpose of process planning. For example the automatic generation of NC code, assembly or quality assurance data require this kind of methods.

  The **functional level**: this involves methods for the generation of the most abstract level of information and includes the generation of calculation models such as for FEM. However, these methods also include the recognition of the function of engineering components. Methods of this type are therefore the basis for analysing solutions or generating libraries of solutions (provided that the solutions are not yet modelled by functional modelling methods).

Figure 3.2-2 illustrates the modelling matrix by providing an example for every field of the matrix.

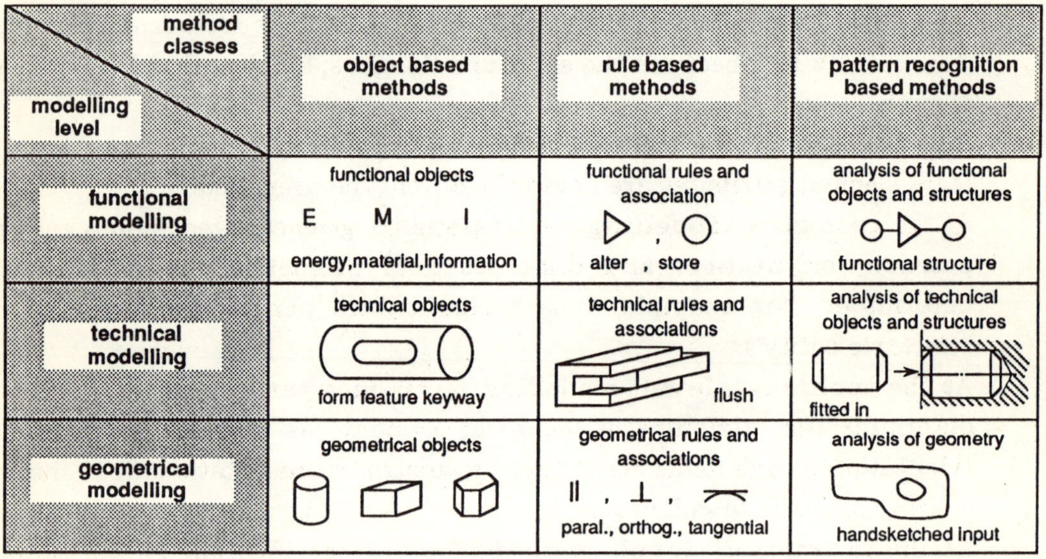

Figure 3.2-2: Examples for the Concepts in the Modelling Matrix

Object based methods are well understood at the geometric modelling level and should in principle also be applicable on the higher modelling levels. However, optimal data structures and methods for technical modelling are still under worldwide discussion. Furthermore, methods for the support of functional modelling are themselves still in a very early stage of research /GRB-89/, /BEN-90/.

The second column of the modelling matrix is based on associative modelling methods. It has been shown that the availability of these kinds of methods is essential for intelligent modelling /GRR-88b/.The major research effort of the CAD*I project on Advanced Modelling was directed at the geometric and technical modelling levels. In particular, the object and rule based methods were dealt within the DTT (Design by Technical Terms) Modeller.

The third column of the matrices also in a very early stage of research. In this column the CAD*I project focused on the geometric level with the system for the input of freehand sketches,the so called Handsketched Input System (HIS). Approaches for the analysis of geometry and for the recognition of technical elements are part of the ESPRIT project "IMPPACT" /IMP-89/.

3.3 General Principles of Associative Modelling

The term "associative" was originally introduced by Seiler /SEI-85, p. 139 ff/. His interpretation of the term "associative" is that of a link between different ideas such that one idea causes the other to be recalled by "associativity".

Applying to the term "associative" yields the following definition:

Associative modelling is understood as a collection of modelling techniques, where some of the modelling parameters can be derived from the geometric, technical, or functional context.

Geometric-associative modelling techniques are based on a set of five primitive geometric constraints:

- identity,
- equality,
- parallelism,
- coaxiality and
- angularity.

However, Seiler did not provide a formal proof that this set of constraints is necessary or complete.

For this reason a different approach has been evaluated. This involves a slight change in the definition of the term "associative modelling". In addition to the definition of Seiler, the geometric constraints are not commutative but that they are directed and one has to distinguish the order of the operands: Although at first sight there is no difference between for example the constraints "A is-parallel-to B" and "B is-parallel-to A", this is nevertheless not so, because the order of operands involves different modifications within the product model.

In a wider context such constraints will cause sequences of effects on entities of a model. Algorithms for propagating changes must consider chains of constraints. As a result the new definition is:

Associative modelling techniques allow information subsets of a product model to be derived from relations to already existing parts of the model.

The term relation is to be understood in the sense of formal logic. Depending on the number of things related to each other, binary, threeary or higher arity relations are needed. In this project only binary relations have been explored. These relations cannot always be assumed to be commutative. When A is-the-father-of B it is obvious that B is-the-father-of A is wrong. However,when A is-parallel-to B, then B is-parallel-to A is true.

Therefore, the order of the arguments linked by a binary relation is important. The first argument is called the action element the second one the reference element.

A classification of constraints applied to the entities of a geometric model is now possible.

The modeller used in the project allows single vertices, edges, sets of edges, faces, sets of faces, as well as volumetric b-rep type objects to be handled. Planar, conical and cylindrical surfaces are available in the system. The line types supported by the system are straight lines, circular lines, elliptic and general section curves. Typical objects which can be modelled based on this model schema are represented in Figure 3.3-1.

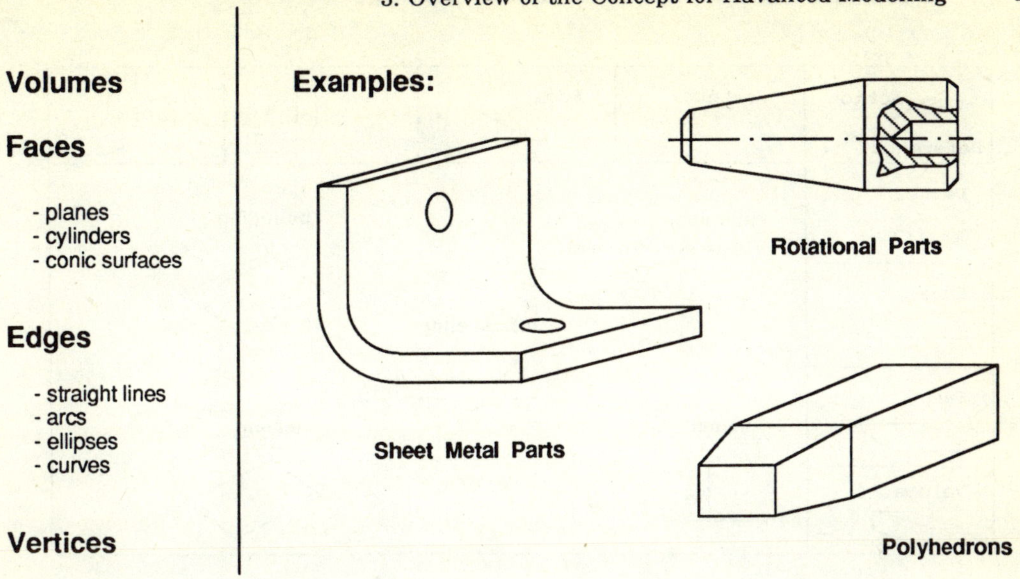

Figure 3.3-1: Typical Objects of the Modeller used in the Project

Based on these types of geometric elements, the possible relationships between them have been explored. A distinction between "trivial" and "interesting" associativities. Trivial associativities are those which are implicitly included in a b-rep schema such as face adjacency. Interesting associativities are defined as those which can provide additional information about geometric elements.

Binary associativities can be classified by means of a matrix of possible relations between two sets of elements:

Action Set = (Vertex, Edge, Face, Volume)
Reference Set = (Vertex, Edge, Face, Volume)

Some of the resulting trivial associativities are illustrated in Figure 3.3-2. The characteristic of trivial associativities is that they contain only low-level information and that they are very numerous.

For example, a vertex can be identical to another vertex, an edge identical to another edge, etc. A vertex can be distant from another vertex, from an edge, from a face, or even from a volume. Similarly edges, faces and volumes can also be distant from other geometric elements. Furthermore, a geometric entity may intersect, bound or be an element of another geometric entity.

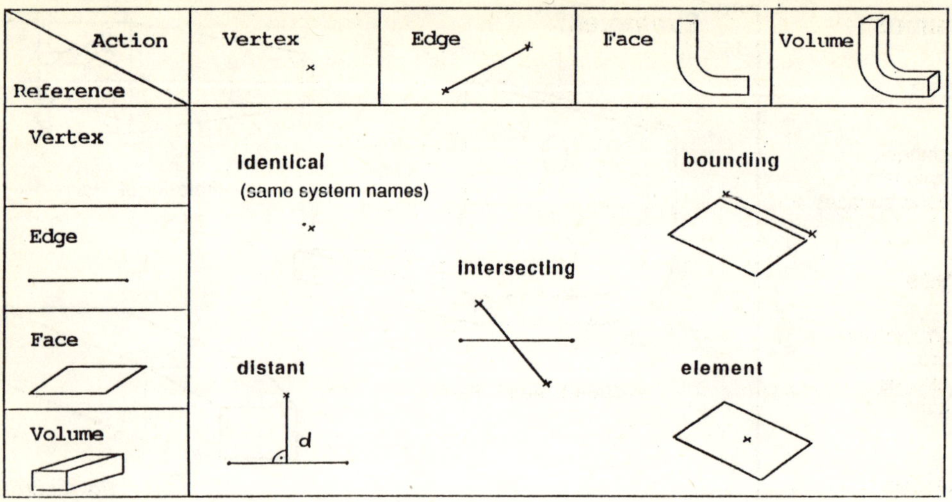

Figure 3.3-2: Examples of "Trivial" Associativities

However, higher-level information can be obtained from considering further possible relationships between action and reference elements. For example, the terms published in former papers e.g. /GAR-86/ are now inserted into the classification matrix (see Figure 3.3-3).

Reference \ Action	Vertex	Edge	Face	Volume
Vertex	(not equal) equal	aligned	aligned	
Edge	aligned	parallel angular aligned concentric	parallel angular coaxial	
Face	aligned	parallel angular coaxial	parallel aligned coaxial angular flush imposed	imposed
Volume			imposed	imposed fitted in

Figure 3.3-3: "Interesting" Associativities

In order to obtain a clear distinction between geometric and technical associativities consider the following definitions

- Geometric associativities are constraints between geometric elements which allow some attributes of the geometric elements to be determined. No analysis of the technical context is needed to obtain an unambiguous result.
- Technical associativities are those associativities whose meaning is determined by looking at the technical context.

If some of the "interesting" associativities can lead to modifications of the geometric elements only if the technical context is clear, then the meaning of "technical context" must be clarified:

"Technical modelling" is defined as modelling technical elements, i.e. products, assemblies, piece parts, form features as well as their attributes (position, orientation, dimensions, tolerances, material, surface quality, etc.). "Technical-associative modelling" is defined as modelling the relations between such technical elements. This information often allows some of the attributes of technical elements to be determined. Both, technical modelling and technical-associative modelling are required to support the preliminary design phase. Technical elements are:

- products
- assemblies
- piece parts
- form features

The action and reference sets must be augmented for the classification of possible relations between technical elements:

Action Set =	((Geometric entities), (Geometric Associativities), Technical elements))
Reference Set =	((Geometric entities), (Geometric Associativities), Technical elements))

Some examples of technical-associative modelling are shown in Figure 3.3-4.

TECHNICAL DESIGN TERM	COMMAND EXAMPLE	APPLICATION EXAMPLE
FITTED IN	INSERT plain pin DIN 6325 (dig 1) INTO bore hole (dig 2) FITTED IN D = H 7, m 6 FLUSH (dig3)	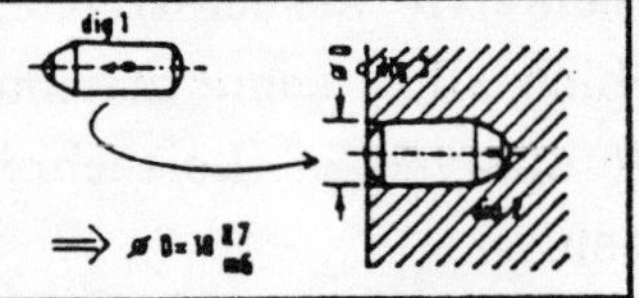
	CREATE die FITTED IN cut - out (dig 1) of die plate (dig 2)	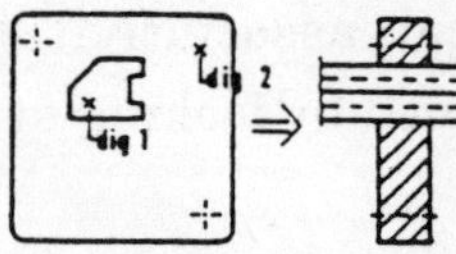
FLUSH	INSERT bar (dig 1) INTO clip (dig 2) , DISTANCE 10 to (dig 3) , FITTED IN (dig 4) , FLUSH WITH (dig 5) , (dig 6)	
	CREATE boundary surface AT die (dig 1) FLUSH TO die plate (dig 2)	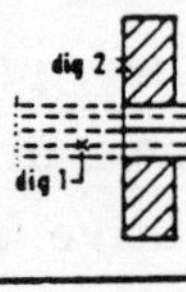
AXIALLY PARALLEL	INSERT gudgeon AXIALLY PARALLEL WITH DISTANCE x TO shaft 1 AND shaft 2	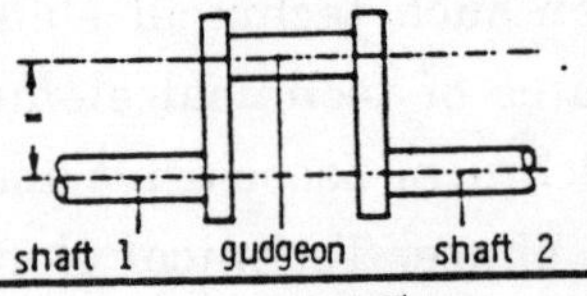
	CREATE cylindrical columns AXIALLY PARALLEL TO cut - out (dig 1) WITH DISTANCE x, POSITION (dig 2) , (dig 3)	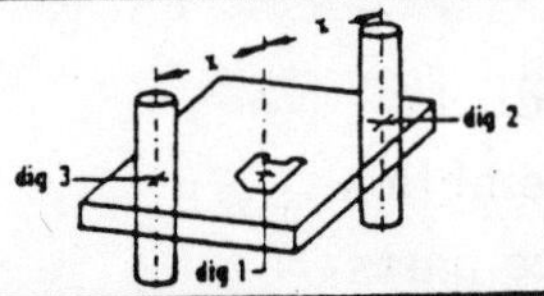
COAXIAL	INSERT shaft segment COAXIAL TO shaft 1	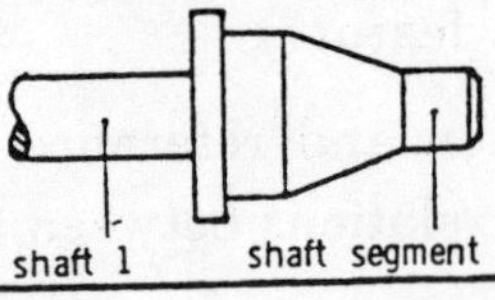
	CREATE stamping head OUT OF MFE shaft, POSITION COAXIAL TO die (dig 1)	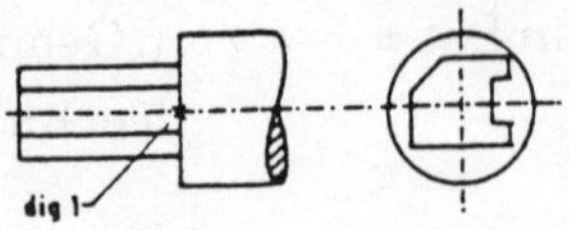

Figure 3.3-4a: Examples of the Use of Technical-associative Terms (part 1)

TECHNICAL DESIGN TERM	COMMAND EXAMPLE	APPLICATION EXAMPLE
FLANGE - MOUNTED	INSERT flange AND seal AT casing FLANGE - MOUNTED BY 4 screws M 8 x 20 DIN 931 IN circular gap	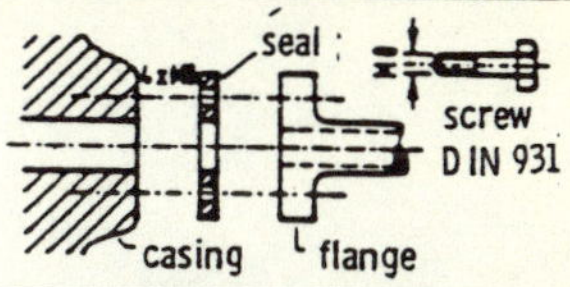
	CREATE connection TO casing FLANGE - MOUNTED AT (dig 1)	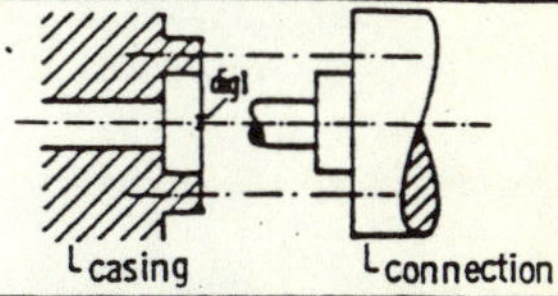
PLAIN PARALLEL	INSERT wipe PLAIN PARALLEL WITH DISTANCE a FROM base plate TO top plate	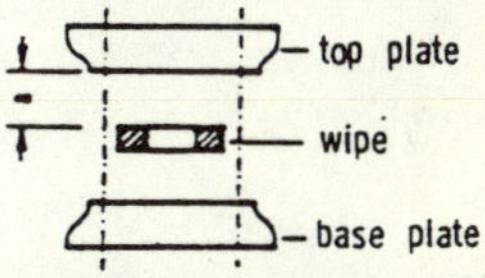
	CREATE boundary surface AT die FITTED IN AND PLAIN PARALLEL TO (dig 1) WITH DISTANCE a	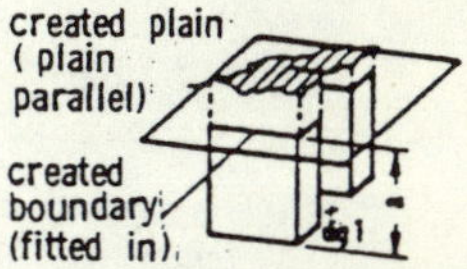
ALIGNED	INSERT support 1 (dig 1), dog (dig 2), support 2 (dig 3) ONTO base plate ALIGNED TO straight line (dig 4), (dig 5)	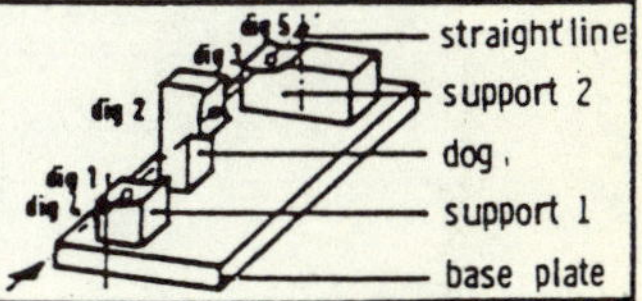
	CREATE shaft ALIGNED TO support 1 (dig 1) AND support 2 (dig 2)	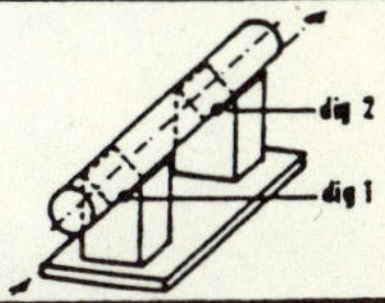
COMBING	INSERT transmission shaft 2 COMBING WITH transmission shaft 1 AND main shaft	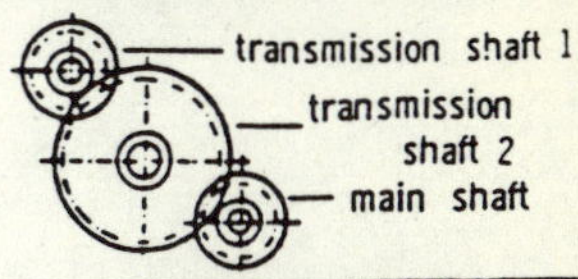
	CREATE gear COMBING WITH main shaft AND output drive shaft	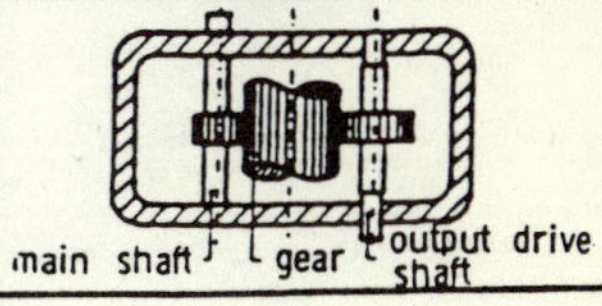

Figure 3.3-4b: Examples of the Use of Technical-associative Terms (part 2)

Further details are described in chapter 4 (geometric-associative) and chapter 5 (technical-associative).

4. Geometric-Associative Modelling

Bernd Pätzold

4.1 Subject of Specification

The goal of an advanced CAD-system is to provide a tool box of automated problem solving aids that allows a designer to conceive, evolve and document his design ideas. Above, it has been described a multi-level advanced modelling system. Because of the important role of geometric information in design, a requirement for an intelligent CAD-system is to store in an explicit form geometric knowledge in a model. In the following a paradigm for an advanced CAD-system is discussed based on the geometric-associative modelling approach.

The main aspects of geometric-associative modelling are:

- the definition and use of geometric constraints,
- an extended user interface based on geometric associative communication techniques,
- enhanced modification operations.

Definition and Use of Geometric Constraints

Design users have to communicate with current CAD-systems on a very low level in terms of geometric operations or primitives. Typical examples are the definition of a line by two points, the intersection of two curves, or more complicated functions such as blending or filleting surfaces, performing a boolean operation on solids, etc. The entities manipulated by the operations must be explicity and fully defined beforehand. Such systems cannot tolerate partial information. The user has to define the entity (point, curve, surface, etc.) and must know precisely all the attributes (coordinates, length. orientation etc.) that uniquely define the entity in this context.

The system must support the definition and use of geometric constraints so as to increase its "knowledge" on the geometric level. Geometric constraints describe relationships between the geometric entities of parts, form features or even geometric primitives:

Extended User Interface based on Geometric Associative Communication Techniques

All user input to a CAD system is defined by an operator (insert, modify, delete, etc.), its operands (point, line, group, etc.) and attributes. One approach to intelligent CAD-systems involves substituting such basic commands with the definition or recognition of geometric associativities. Possible methods for defining geometric associativities are:

- defined explicitly by a user command
  (e.g. Make line parallel...)
- defined implicitly by
  - the semantics of a user command (e.g. Insert Line parallel)
  or
  - in terms of form features (e.g. fillet, chamfer, etc.)
- defined in consequence of the recognition of the geometric context.

Enhanced Modification Operations

Today's CAD systems are not very effective in supporting the modification of existing models or drawings because the metric relationships between geometric entities are not part of the CAD model. The availability of geometric associativities are a preliminary requirement for the effective support of the designer.

A geometric-associative modelling system must satisfy the following minimum of requirements:

- augment the geometric CAD model by geometric associativities
- allow geometric associativities to be defined through both conventional and more advanced communication techniques
- provide the user with appropriate feedback at every step of the design
- provide enhanced modification operators.

4.2 Geometric-Associative Modelling Approaches

Alternative approaches to mechanical engineering design will now be discussed in terms of parametric modelling. These approaches all have in common that they augment conventional geometric modelling by the inclusion of geometric associativities.

There are three basic alternatives /RSV–89/:

- Constructive Approaches,
  based on algorithmic solutions.
- Design Language Approaches,
  based on a language for the definition and manipulation of parametric objects.
- Rule Based Approaches,
  based on artificial intelligence methods.

4.2.1 Constructive Approaches

A characteristic of constructive approaches is that the types of geometric associativities are limited and that the derivation of a solution requires well know algorithms and may involve solving a system of equations.

Early concepts like those of /HIB-78/ or /FIT-81/, /KDY-85/ are based on systems of geometric associativities between the vertices of a geometric model. These vertex oriented descriptions define a system of equations based on the distances between vertices, angular conditions and distances between parallel lines. The solution of the system of equations is mainly based on the Newton-Raphson method. Figure 4.2.1-1 shows a dimensioned polygon and its corresponding set of constraints, called "stiffener" by /HIB-78/.

A system of non-linear equations can be derived from these constraints. The constraint of a 2D part requires a number of drafting conventions. Figure 4.2.1-2 shows such a set of drafting conventions together with their corresponding constraint equations as definded by /LIG-82/.

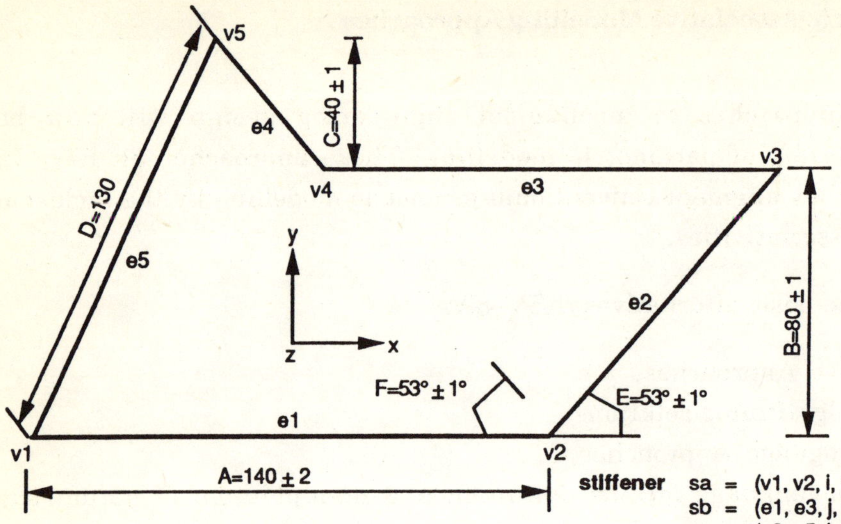

stiffener sa = (v1, v2, i, 140, 2, 1, 0, 0)
sb = (e1, e3, j, 80, 1, 1, 1, 0)
sc = (e3, v5, j, 40, 1, 1, 0, 0)
sd = (v1, v5, 5*j+12*j, 130, 0, 1, 0, 0)
se = (e1, e2, k, arctan(4, 3), 1, 0, 1, 0)
sf = (e1, e4, k, arctan(4,-3), 1, 0, 1, 0)

Figure 4.2.1-1: A Dimensioned Polygon and Geometric Constraints /HIB-78/

Dimension name	Entities constrained	Equation
Horizontal distance	P_1, P_2	$X_1 - X_2 - D = 0$
Vertical distance	P_1, P_2	$Y_1 - Y_2 - D = 0$
Linear distance	P_1, P_2	$(X_1 - X_2)^2 + (Y_1 - Y_2)^2 - D^2 = 0$
Distance from point to line	$P_1, \overrightarrow{P_2 P_3}$	$\hat{U} \times V - D = 0$ $\hat{U} = \frac{(X_3 - X_2)}{\lvert\overrightarrow{P_2 P_3}\rvert} i + \frac{(Y_3 - Y_2)}{\lvert\overrightarrow{P_2 P_3}\rvert} j$ $V = (X_2 - X_1) i + (Y_2 - Y_1) j$
Angular dimension	$\overrightarrow{P_1 P_2}, \overrightarrow{P_3 P_4}$	$\lvert\overrightarrow{P_1 P_2} \times \overrightarrow{P_3 P_4}\rvert / (\overrightarrow{P_1 P_2} \bullet \overrightarrow{P_3 P_4})$ $- \mathrm{Tan}\ (A) = 0$

Figure 4.2.1-2: List of Dimensional Constraint Equations /LIG-82/

Other approaches /TUW-88/ /GZS-88/ /HOF-87/ are based on the definition of a graph of geometric primitives and constraints. The geometric constraints are represented by relative position or dimension operators or by single scalar parameters. This allows the description of a part in terms of an object graph containing nodes, branches and loops. The nodes of the graph represent operators and the branches describe the sequence of operations needed to compute the b-rep object. Figure 4.2.1-3 shows a complex object graph of a cube with a slot /GZS-88/.

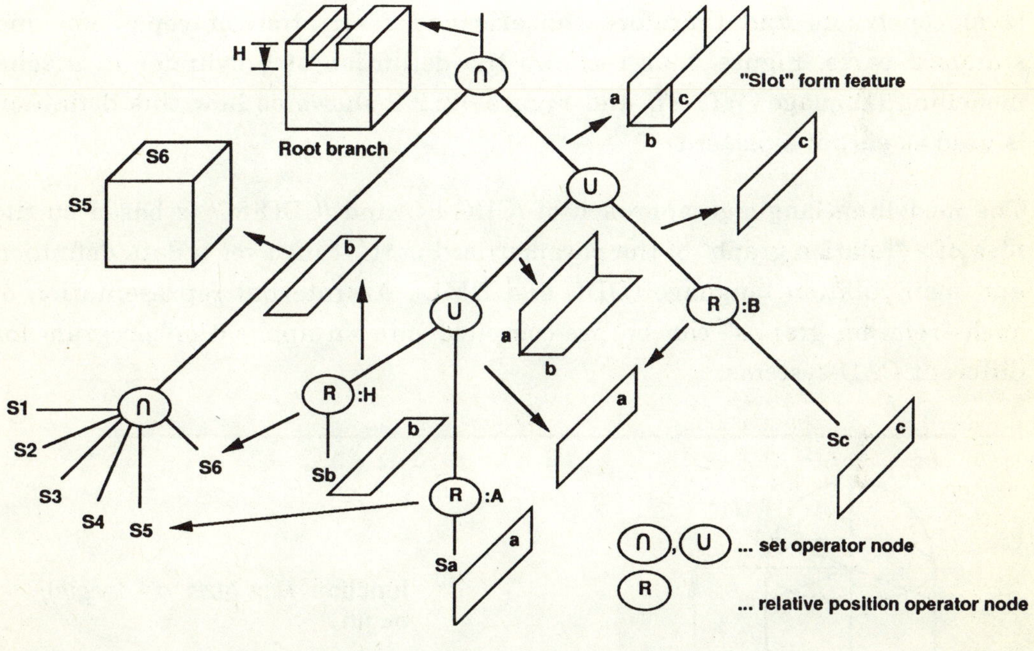

Figure 4.2.1-3: Constrained Object Graph of a Cube with a Slot /GZS–88/

4.2.2 Language Approaches

The language oriented approaches to the definition of objects /WIJ-87/ /CDG-84/ /CDG-85/ are quite different. The user has to prepare a list of alpha-numerical commands which describe the model. This description is then used to generate a program which may require some further user input in order to compute the model of an object. Therefore, language oriented approaches require an a priori abstract definition of a model.

The main advantages of this approach are that the language supports high-level constructs and therefore the efficient description of repetition and standard parts. Figure 4.2.2-1 shows the definition of a cylinder in a solid modelling language /WIJ-87/ and Figure 4.2.2-2 shows as how this definition is used as part of a pattern.

The modelling language approach of /CDG-84/ and /CDG-85/ is based on the idea of a "relation graph" of the parametrised part. It involves a data definition and manipulation language (DDL and DML). An internal representation of such "relation graphs" can be pre-compiled into an application program for different CAD-systems.

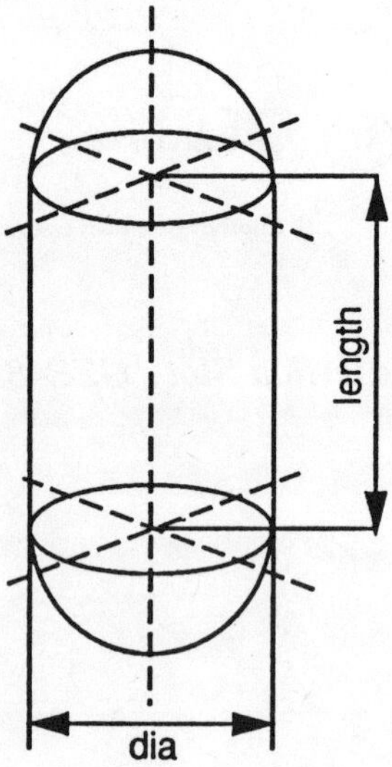


```
function Rcyl (axis, dia, length);
begin
    Rcyl := cyl dia length
    Top  := sph dia
    Rcyl := Rcyl + Top
    Take Top; mov z length
    Rcyl := Rcyl + Top
    if axis = x then rot y 90 else
    if axis = y then rot x -90
end; {Rcyl}
```

Figure 4.2.2-1: Definition of a Segment using a Solid Modelling Language /WIJ-87/

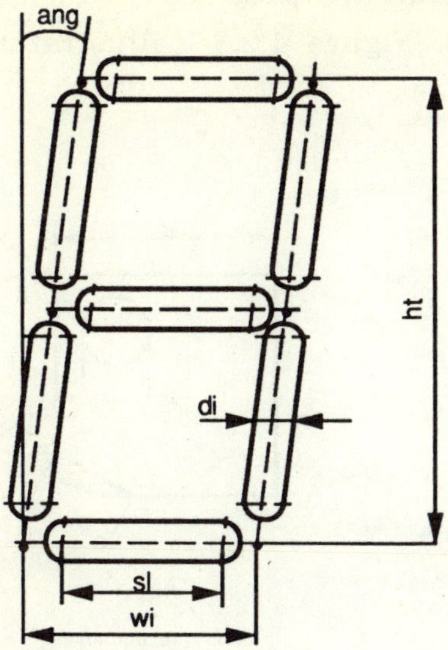

Def Segments(10, 5, 8.5, 1, 8);

```
global Seg : list of object;

procedure Def Segments (ht, wi, sl, dl, ang );
beginn
    hh := ht / 2;
        Seg [ 1 ] := Rcyl (x , dl , sl );
            mov x (wl - sl)/2;
        Seg [ 2 ] := Seg [ 1 ];
            mov y hh;
        Seg [ 3 ] := Seg [ 2 ];
            mov y hh;

    { vertical segments }

        Seg [ 4 ] := Rcyl (x , dl , sl );
            mov x (hh - sl)/2;
        Seg [ 5 ] := Seg [ 4 ];
            mov x wi;
        Seg [ 6 ] := Seg [ 5 ];
            mov y hh;
        Seg [ 7 ] := Seg [ 6 ];
            mov y wi;

        for i := 1 to ! Seg do
        begin
            take Seg [ i ] ; ske y x ang
        end

end; { Def Segments }
```

Figure 4.2.2-2: Definition of a Pattern of Segments using a Solid Modelling Language.

4.2.3 Rule Based Approaches

Another kind of approach for parametric design involves the use of rule based methods of artificial intelligence. In such an approach parametric dependencies are commonly described in terms of rules.

In his "Paradigm for Intelligent CAD" Arbab /ARB-87/ describes a design process which involves the explicit definition of geometric constraints. For geometric reasoning a kind of logic programming approach is used which requires additional knowledge of abstract geometric concepts represented explicitly in predicate logic. Detailed descriptions either the required rules or the integration into existing CAD system architectures are not given.

A more practical approach is given by Sunde /SUN-86/ /SUN-87/ /BJO-87/ and by Aldefeld /ALD-88/.

With the PICME SKETCH-solution Sunde uses an approach based on a "post definition process" for geometric constraints. Figure 4.2.3-1 illustrates a design process with the PICME-system.

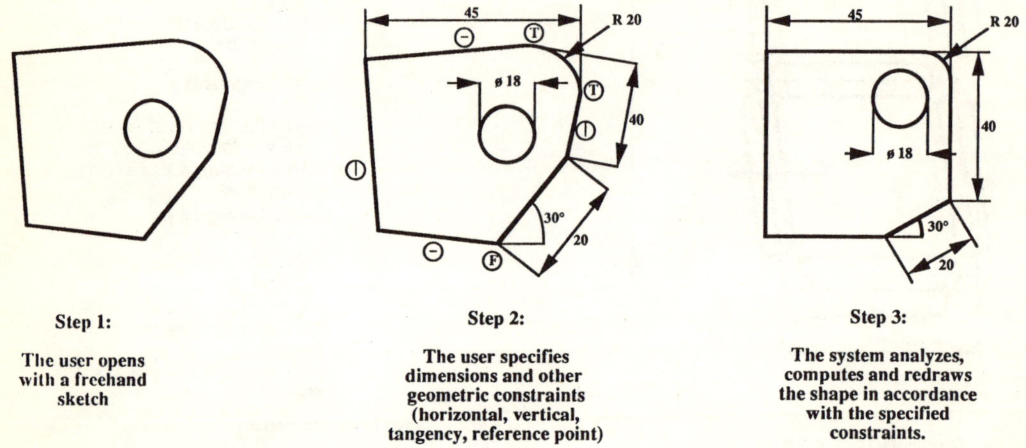

Figure 4.2.3-1: The PICME SKETCH System /BJO-87/

The procedure is based on a freehand sketch input. The actual shape and size are determined by explicit geometric constraints. These constraints are translated into rules for an expert system. The deduction of an accurate geometric model is done automatically by the system.

Experience with the PICME system is available in the area of product development and in particular for the generation of drawings.

A second approach is based on on-line processing of geometric constraints and interactive updating of the current shape definition. Figure 4.2.3-2 shows an interaction sequence of a design process. The process evaluates the design at every step, recognizes overdetermined cases and allows any constraint to be modified during the design process.

The approach described by Aldefeld /ALD-88/ is also based on an expert system. The handling of constraints is similar to that of the PICME system (Figure 4.2.3-3). The computing process is subdivided in two phases:

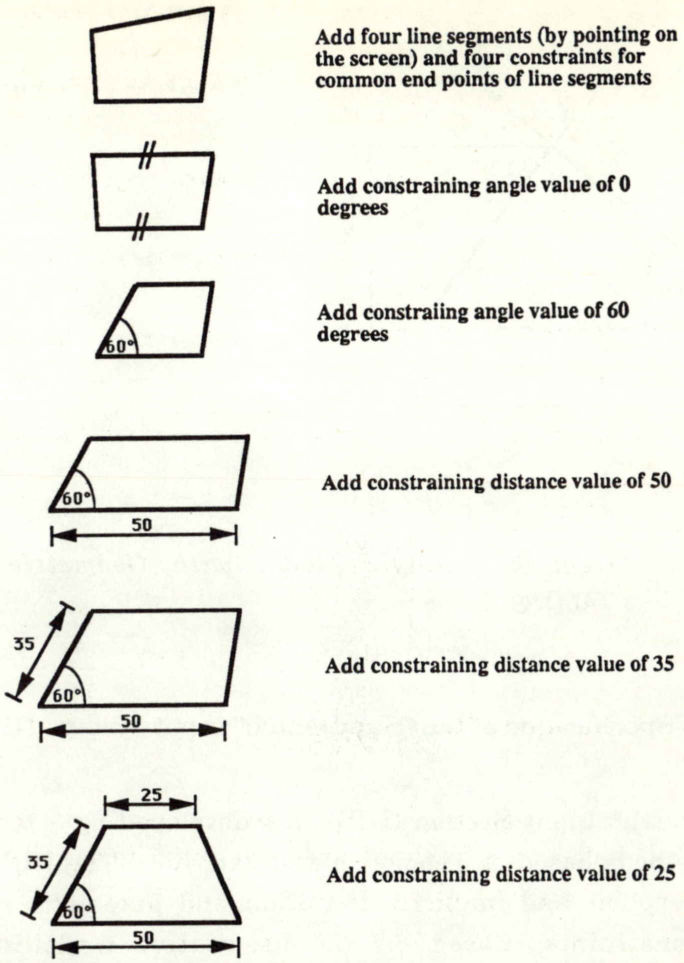

Figure 4.2.3-2: *Example of the Interactive Definition of Geometric Constraints /BJO-87/*

- preprocessing of the design and building a symbolic description
- computation of the numeric values of the design.

The main difference to the approach of Sunde is that he is able to classify constraints as global or area oriented constraints. The inference process is expensive in terms of computation times. For example the system requires 10 minutes of CPU time to compute a part with only 300 points.

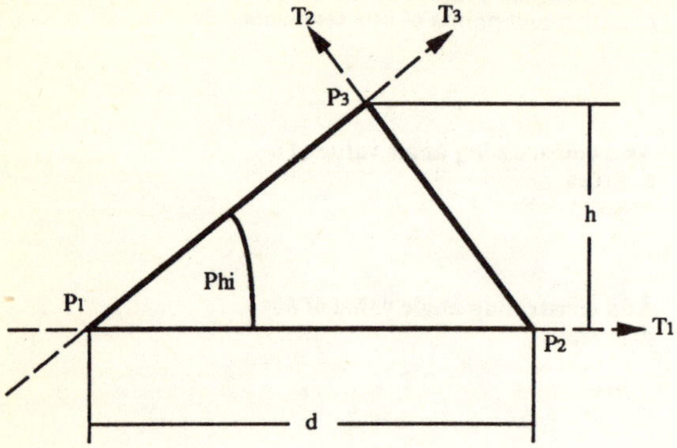

(L-Coeffs T_1 *)	Inferred	Using Rule-D1 (Coords P_1 (X_1 Y_1)) (DirectionT_1 0.0)
(Coords P_2 *)	Inferred	Using Rule-C1 (Coords P_1 (X_1 Y_1)) (Direction T_1 0.0) (Distance P_1 P_2 d)
(Direction T_3 *)	Inferred	Using Rule-E1 (Direction T_1 0.0) (Angle T_1 T_2 Phi)
(L-Coeffs T_3 *)	Inferred	Using Rule-D1 (Coords P_1 (X_1 Y_1)) (Direction T_3 *)
(Coords P_3 *)	Inferred	Using Rule-C4 (L-Coeffs T_3 *) (L-Coeffs T_1 *) (Distance P_3 T_1 h)
(L-Coeffs T_2 *)	Inferred	Using Rule-D2 (Coodrs P_3 *) (Coords P_2 *)

Figure 4.2.3-3: Generic Triangle Model with Geometric Constraints /ALD-88/

4.3 System Specification of the "Handsketch" Input System (HIS)

The "Handsketch" Input System (HIS) was developed as a testbed for new modelling methods based on geometric associativities. The HIS supports the interactive (explicit and implicit) definition and automatic recognition of geometric constraints. Based on the associative modelling approach, modification of a part is well supported.

The HIS provides a new communication technique for the description of geometric or symbolic entities in terms of the input of freehand sketches. It is a user oriented communication technique which is close to conventional design methods.

The main tasks of the "Handsketch" Input System can be described as follows:

1. The user input like sketched geometric entities or geometric symbols have to be recognized on a syntactic level.
2. Any user input has additional semantics depending on the current state of model. It must therefore be interpreted in connection with this semantic context.

3. Any syntactic entities (e.g. geometric entities or symbols) that are recognized should be stored in the model so as to be available for further analysis.

In order to meet these demands the following architecture of the "Handsketch" input process was developed (Figure 4.3-1). The individual modules of the HIS perform the tasks described in the following chapters.

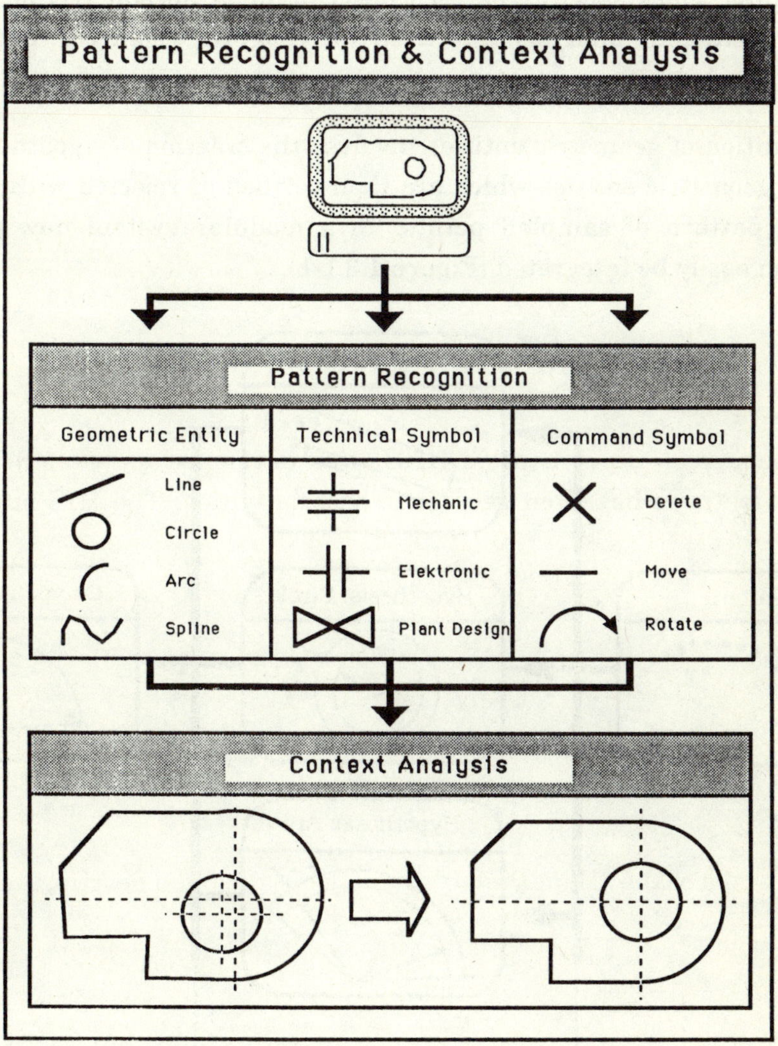

Figure 4.3-1: Architecture of the HIS

4.3.1 The Recognition Process

The recognition process of the HIS can be subdivided into two parts:

- Identification of individual syntactic entities using pattern recognition techniques, and
- Context sensitive analysis of the syntactic entities.

The pattern recognition process examines a given pattern and selects a single, corresponding, class specific, pattern description. A special technique which enables an efficient analysis was used to recognize geometric entities and geometric symbols.

The recognition of geometric entities involves the creation of hypotheses based on known geometric entities which are then verified or rejected with respect to the input pattern of sampled points. In a modular system new geometric entities can easily be integrated (Figure 4.3.1-1).

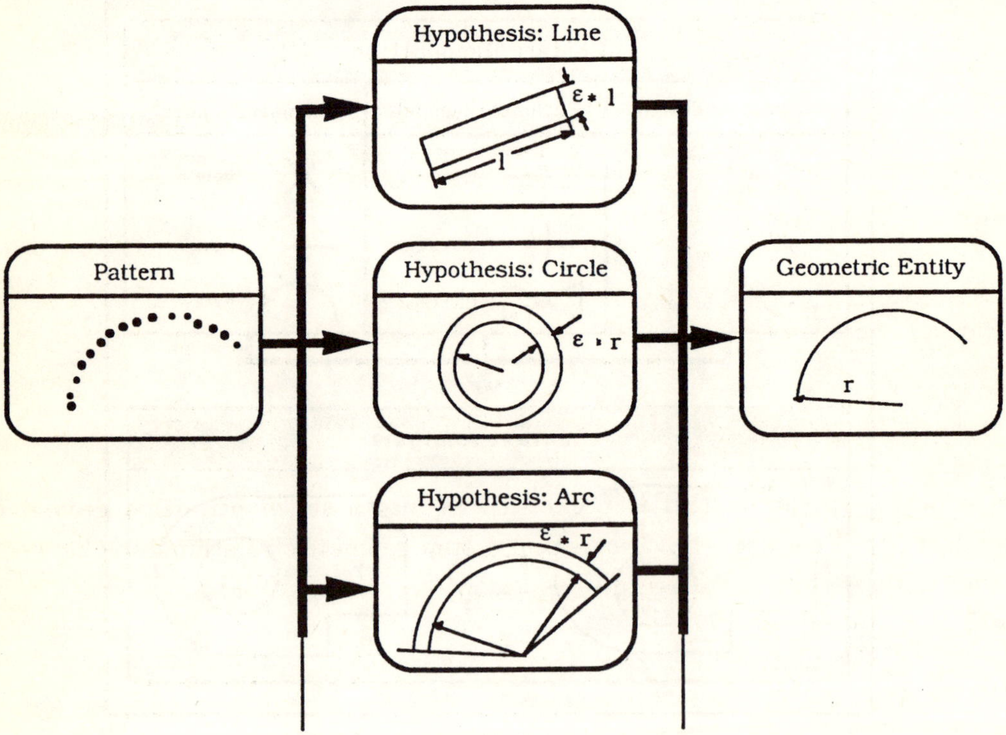

Figure 4.3.1-1: Pattern Recognition Technique for Geometric Entities

The result of the pattern recognition process is a set of parameters describing the type (line, circle, arc, etc.) position, orientation and dimension of the recognized geometric entity or symbol. This information will hereafter be refered to as position, orientation and dimension data (POD).

The POD data is passed to the context analysis system for further processing of the semantic context.

The recognition of both technical and command symbols requires a size and position invariant description of the symbol. Picture Description Languages (PDL) /NAR-68/ are a technique well suited to this task.

These methods involve decomposing an object into geometric primitives, whose relationships and characteristics are then described. Symbol recognition techniques based on such PDL's were used in the GEOMAP /HOK-77/ and CASUS /JNR-85, SKJ-87, KJT-88/ systems.

Pattern Recognition System for symbols are often based on the kind of PDL methods described in Figure 4.3.1-2.

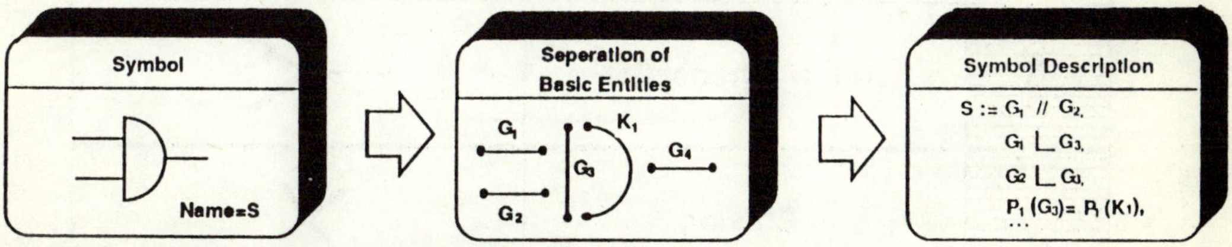

Figure 4.3.1-2: Description of a Symbol using a PDL

In a first step the symbol is broken down into a set of primitive geometric entities (e.g. line, arc, etc.). In a second step geometric relationships between these entities (e.g. parallelism, angularity, etc.) are determined.

A transformation function is then used to obtain a size and position invariant symbol description. The result is stored using the description language and is henceforth available for comparison with other symbols.

Such symbol definitions can be used in two ways

- either to recall complex geometric models (i.e. as a macro function), or
- to initiate certain modelling operations.

On sucessful recognition of a symbol the corresponding macro function or modelling operation is executed (see Figure 4.3.1-3).

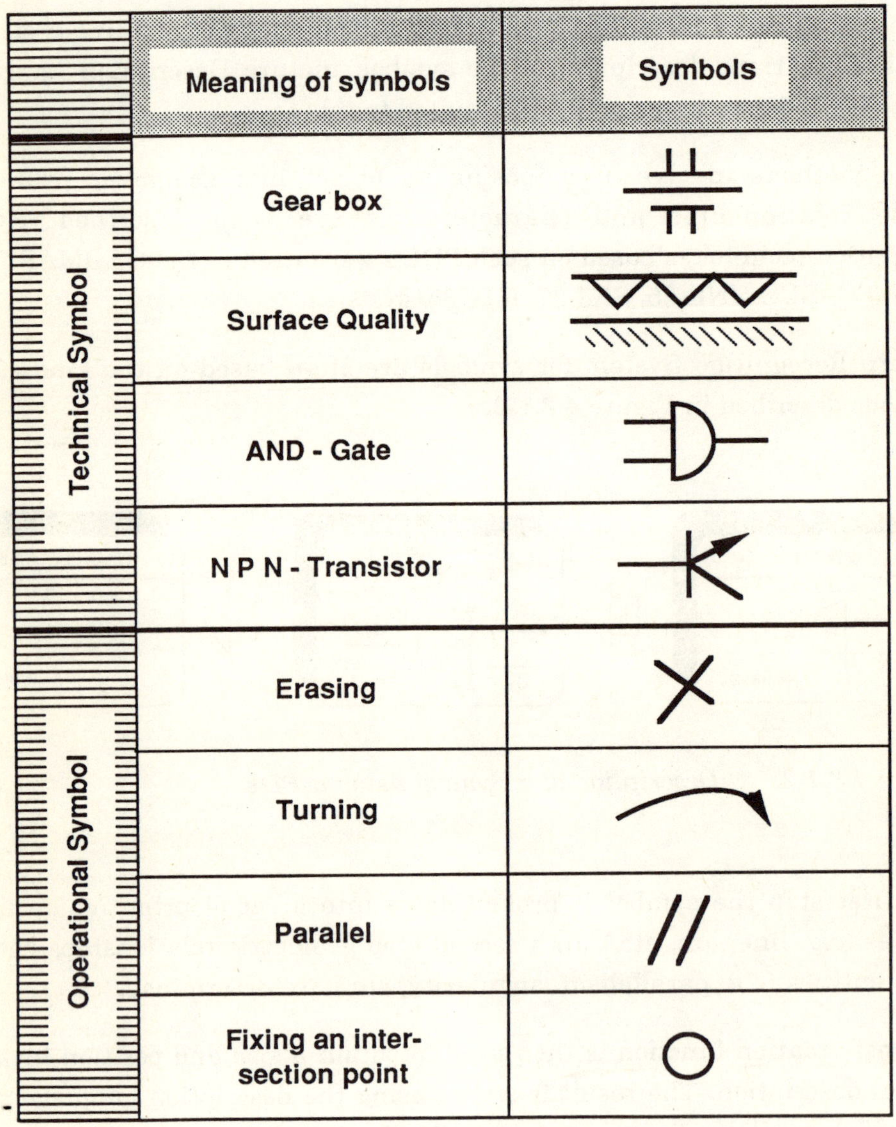

Figure 4.3.1-3: Examples for Technical and Operational Symbols

The context analysis process adjusts the recognized object (geometry or technical entity) with respect to the semantics of the current context before transferring it to the CAD model. This process involves recognizing, retrieving and using geometric relationships such as parallel, angular, co-axial, symmetry. The POD-parameters of a recognized object are used to correct its geometric description that a semantically meaningful adjustment will be performed (Figure 4.3.1-4)

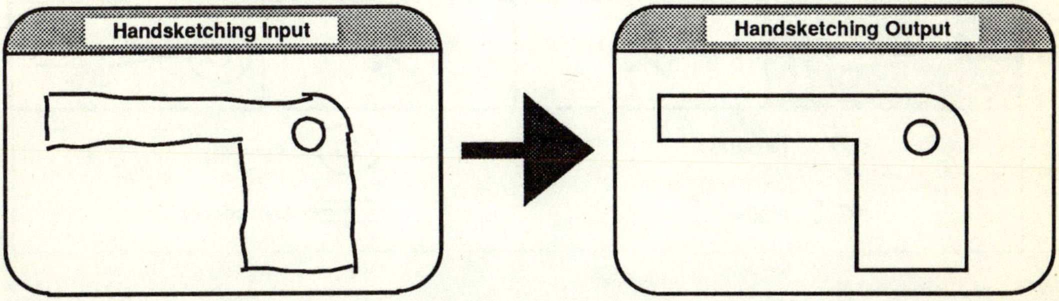

Figure 4.3.1-4: *Example of a Context Based Semantic Adjustment of a Freehand Sketch*

4.3.2 The Rule Base

In the proposed context analysis process the relevant geometric relationship is determined by the geometric context. A set of possible geometric relationships must therefore be assigned to each possible pair of geometric entity types (Figure 4.3.2-1).

The geometric relationships are:

- Parallel
- Angular
- Co-Axial
- Symmetric

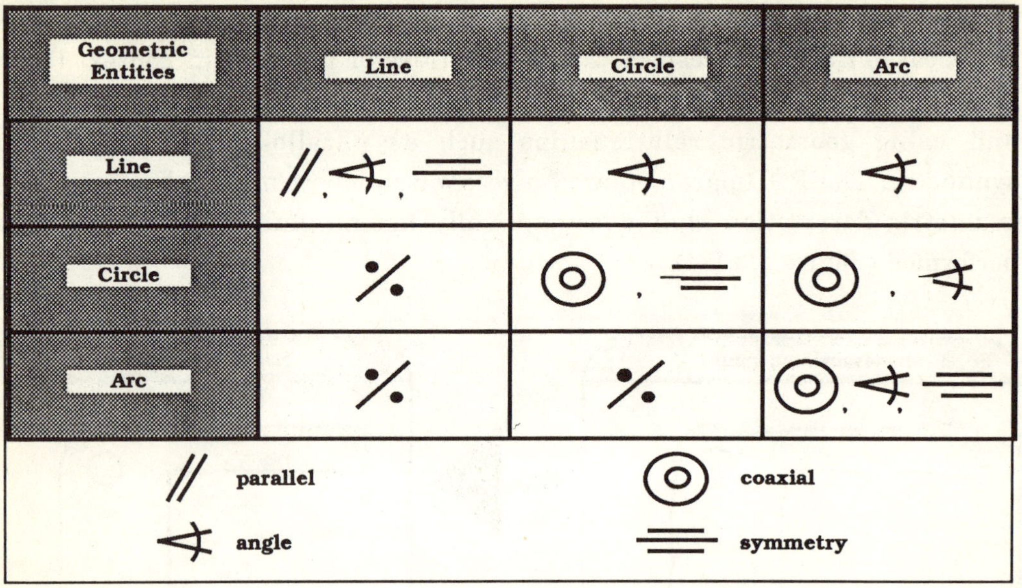

Figure 4.3.2-1: Assignment of Geometric Relationships to Pairs of Geometric Entity Types.

Depending on the kind of entity types, a given geometric relationship will have different semantics. These semantics are described in terms of rules so that they can be executed when needed. Figure 4.3.2-2 shows an extract of the rule base.

Every rule consists of a condition and an action part. The condition part describes the requirements which have to be fulfilled so that the rule can be applied. The action part defines a set of actions (operations) which will be performed to change the model. Because of the modular structure of the rule base, it is possible to integrate any CAD modelling interface.

Any object can be described by a set of geometric entities (themselves described in terms of their type and POD parameters) and by a set of relationships between these entities. Geometric relationships are considered as constraints between geometric entities and their POD parameters.

Geometric relationships can be either defined explicitly by the user or implicitly by the current geometric context.

GA	Rule	Example
//	par (i, j) := i ∈ typ (LINE), j ∈ typ (LINE), $\|angle (i, j)\| < \zeta$, => ins_par (i, j)] , [. . .] .	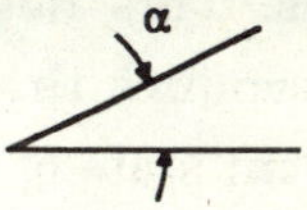 $\|\alpha\| < \zeta$
	ang (i, j) := [i ∈ typ (LINE), j∈ typ (ARC), tangent (j, P, k), $\|angle (i, k)\| \zeta$, => ins_tan (i, j)] , [. . .] .	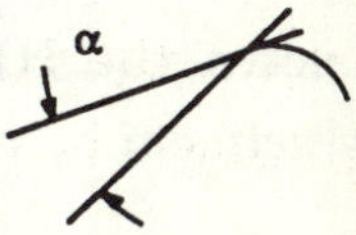 $\|\alpha\| < \zeta$
	coax (i, j) := [i∈ typ (CIRCLE), j∈ typ (CIRCLE), $\|abs(M(i), M(j))\| < \zeta$, => ins_coax (i, j)] , [. . .] .	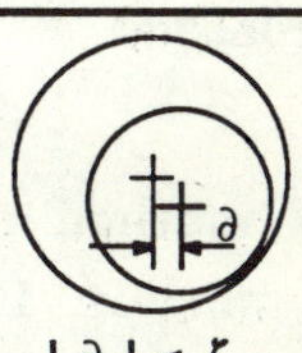 $\|\partial\| < \zeta$
≡	sym (i, j, k) := [i ∈ typ (ARC), j ∈ typ (ARC), k ∈ typ (SYM_LINIE), $\|abs(i, k) - abs(j, k)\| < \zeta$, => ins_sym (i, j, k)] , [. . .] .	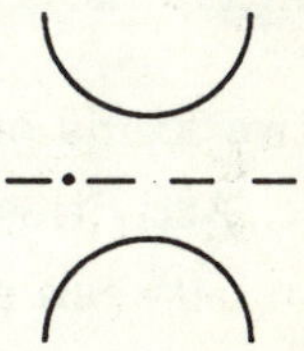
	Legend : i - Existing Entity j - New Entity k - Tangent or symmetry line P, M - Point	

Figure 4.3.2-2: Rule Base of the Semantic Context Analysis System

4.3.3 Architecture of the System for the "Handsketch" Input System

The HIS consists of two independent modules. The first module is the Pattern Recognition System (PRS) responsible for the syntactic recognition of geometric entities and the other the Context Analysis System (CAS).

The PRS examines an input pattern of sampled points (the freehand sketch) and a single corresponding syntactic entity. This entity is described by its type (straight line, circle or arc) and its corresponding position, orientation and dimensions. This description is passed into the Context Analysis Module which determines the semantic relationships (if any) of the recognized entity to the current state of the model.

The structure of the HIS system will be described using the IDEF method /ICA-81/ which is itself a subset of the SADT /ROS-85/. This method involves breaking down the HIS system in successive steps until logical units are reached which can be realized by a set of procedures or functions.

The IDEF Method

The IDEF method exists of a modelling language for the definition of functional models. Planned and existing systems can be modelled using graphical techniques. Extensive system descriptions can therefore be replaced by IDEF-diagrams.

The only two kinds of graphic symbols used by the IDEF-language are boxes and arrows. Activities are represented by boxes. Relationships and the flow of information between activities are defined by arrows. There are three different types of arrows:

INPUT/OUTPUT: An INPUT arrow is used to represent information or material, which will be transformed by the activity into an information flow which will be OUTPUT.

CONTROL: A CONTROL arrow is used to represent information, which influences the function.

MECHANISM: A MECHANISM arrow is used to represent the mechanism which supports an activity and realises the corresponding function:

Figure 4.3.3-1 describes the IDEF-method and illustrates the numbering scheme for the arrows.

The IDEF method describes the structure of a system as a hierarchical diagram. At the top level of the hierarchy the complete system is described as a black-box. This high level function is then successively subdivided in a top-down manner until the required level of detail is reached. Normally, only three or four levels are needed. Figure 4.3.3-2 provides an example of a hierarchical IDEF diagram and illustrates the numbering scheme for the activities.

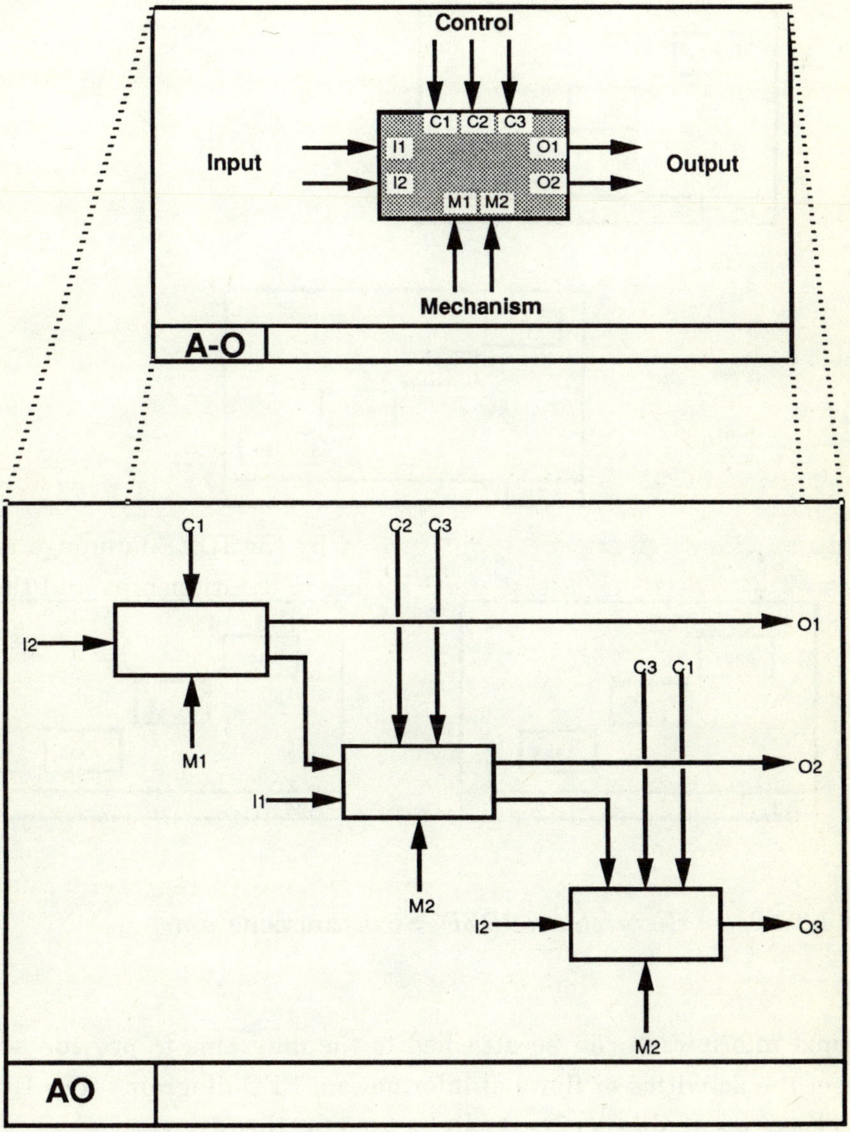

Figure 4.3.3-1: The IDEF-Systematology

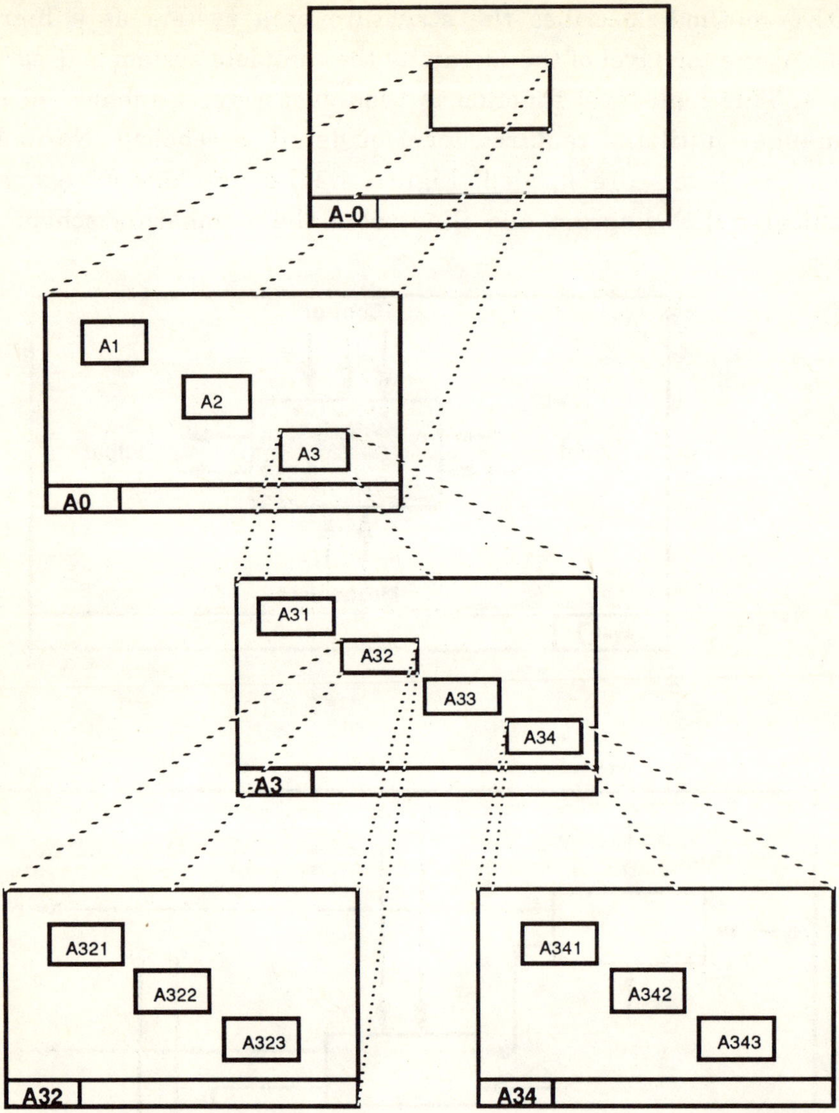

Figure 4.3.3-2: *Hierarchical IDEF Structure Diagram*

Additional information can be attached to the diagrams to provide additional details of the activities or flows of information. FEO-diagrams (For Exposition Only), glossaries or sheets of text can be used for this purpose.

The "Handsketched" Input System (HIS) in IDEF

The HIS is described using the IDEF method and reaches a level of logical units which can be realized directly by sets of procedures or functions.

Figures 4.3.3-3 through 10 give an overview of the system.

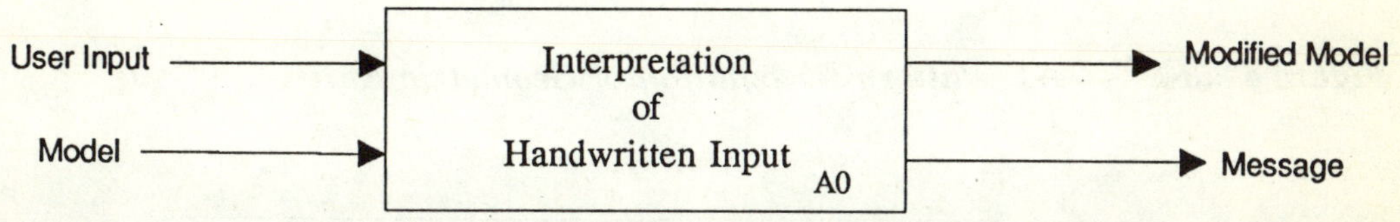

Figure 4.3.3-3 A-0 Handwritten-Input System Architecture

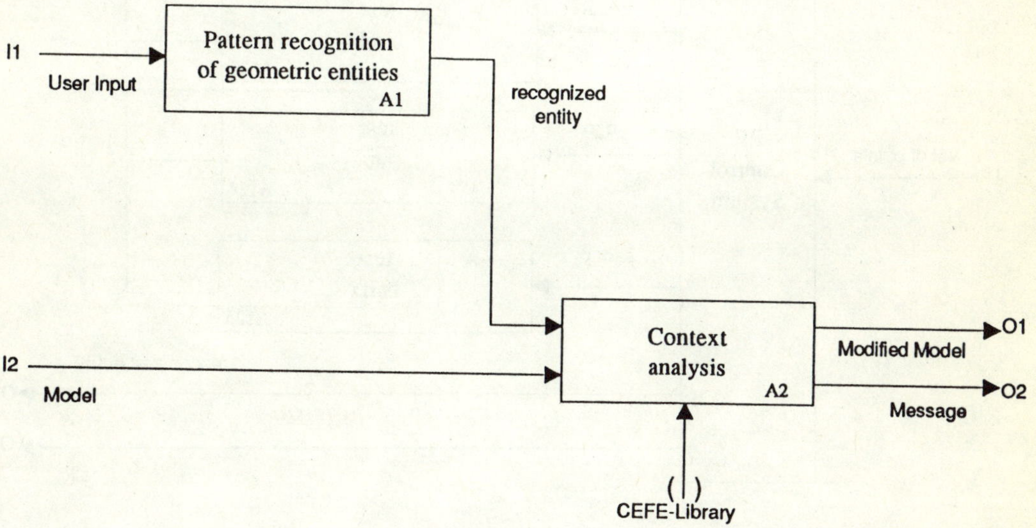

Figure 4.3.3-4 A0 Interpretation of Handwritten Input

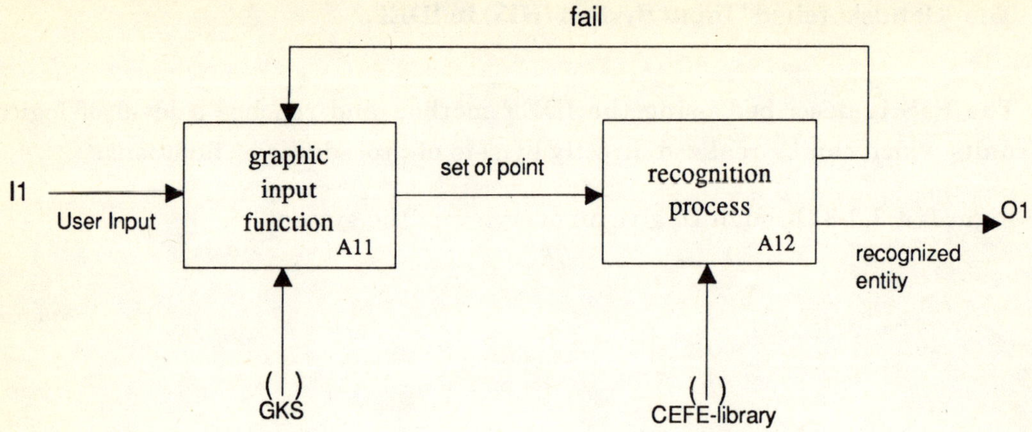

Figure 4.3.3-5 *A1 Pattern Recognition of Geometric Entities*

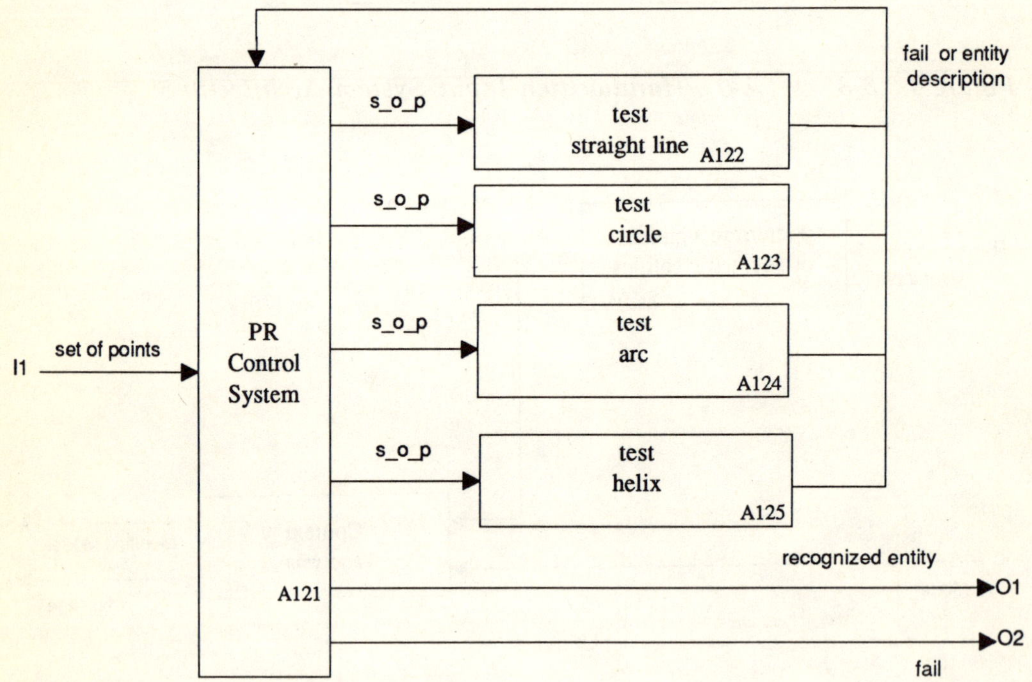

Figure 4.3.3-6 *A12 Recognition Process*

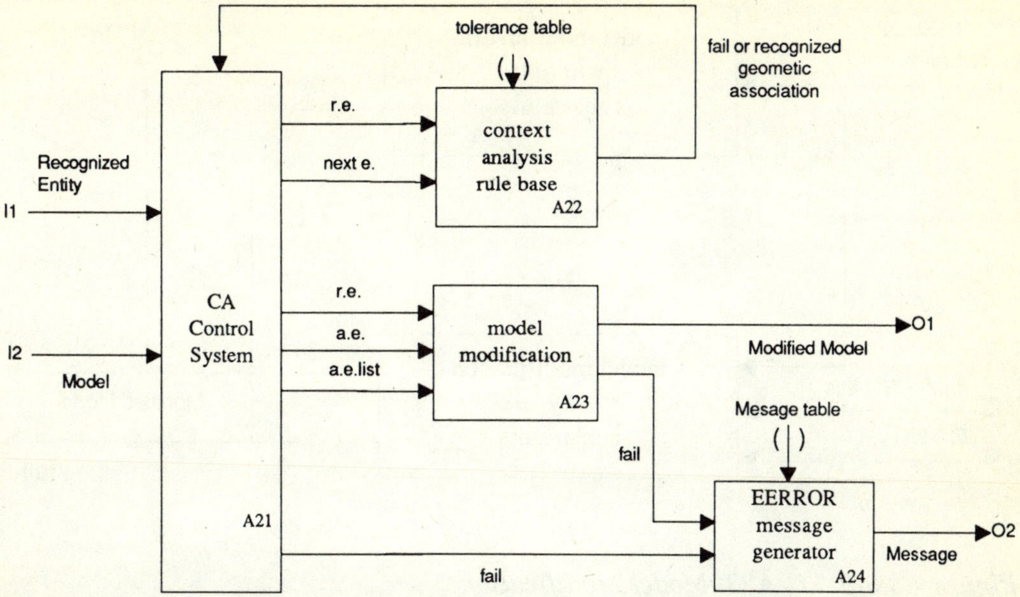

Figure 4.3.3-7 *A2 Context Analysis*

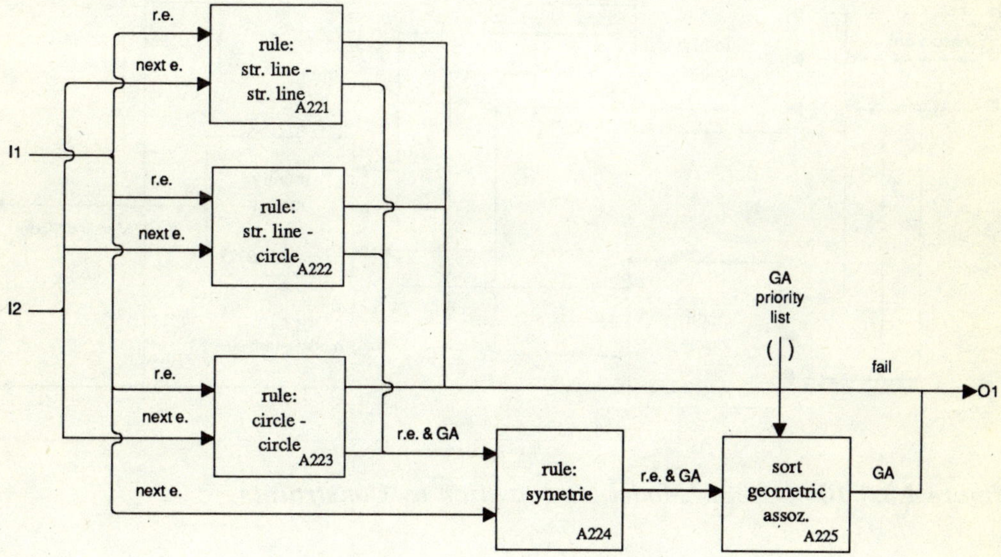

Figure 4.3.3-8 *A22 Context Analysis Rule Base*

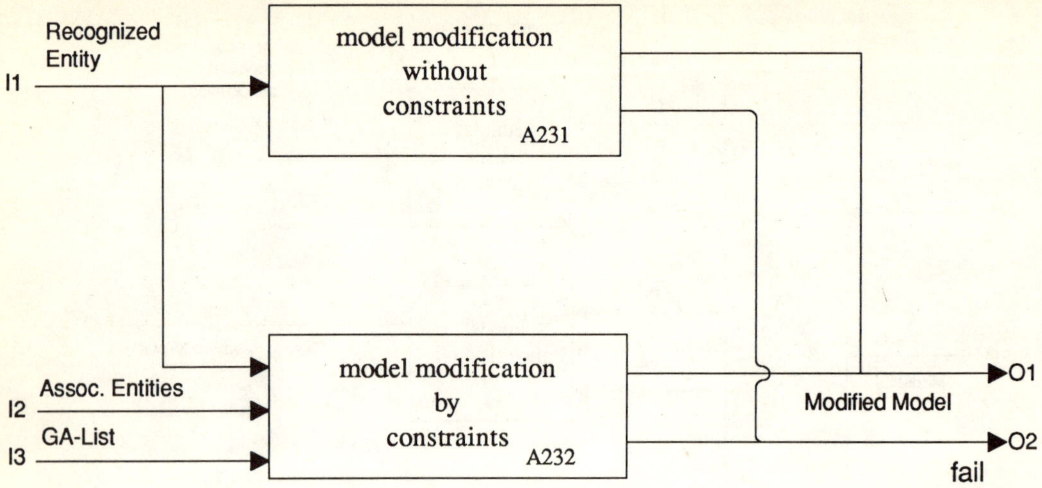

Figure 4.3.3-9 A23 Model Modification

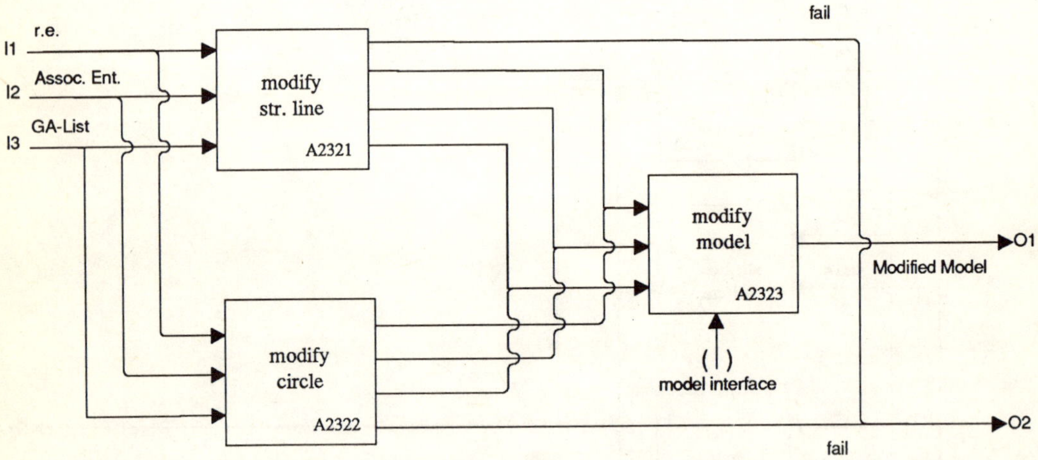

Figure 4.3.3-10 A232 Model Modification by Constraints

5. Technical-Associative Modelling

Stefan Rude

5.1 State of the Art

Since the CAD*I project started in 1984 a number of systems have become available which realize some aspects of the modelling concepts, described in Chapter 3. An overview of systems which, to some extent, include associative modelling techniques is given in Figure 5.1-1.

System :	Advanced Modelling Techniques Used
Cimplex	Design by features
Cognition	Variational geometry
ICAD	Design language
Iconnex	Variational geometry
Intergraph	Associative Modelling
Novocad	Variational geometry
Prime/CV	Variational geometry
PRO/Engineer	Design by features, 3D parametric modelling
Sigraph CAD Blech	Design by features

Figure 5.1-1: Advanced Modelling: List of Implementations

Obviously, it cannot be guaranteed, that the list is complete. The first point is that only 3D-systems have been considered as possible members of the list. The second point is, that currently most commercial systems cover the aspect of assembly modelling at least to some extent. However, in most cases the systems cannot model associativities between parts. Instead, they only provide part-lists like structures.

The more advanced systems in the above list have been explored in greater detail as shown in Figure 5.1-2.

System Name:	ICAD	Pro/ENGINEER	CIMPLEX	INTERGRAPH
Vendor:	ICAD Inc. Cambridge, U.S.	Parametric-Tech. Corporation, U.S.	Cimplex Inc. California, U.S.	Intergraph Corp., U.S.
Classification:	Language based design	Feature-based design parametric 3D, based on NURBS	Feature-based design 3D-solid	Solid modeling based on NURBS
Modelling Techniques:	Product definition with a symbolic, objectorientated language, design rules	Features defined by 2D-input and sweep operation, 3D-parametrics	Design by features	Associative modelling techniques available for free form surfaces
Remarks:	Only simple primitives available, unevaluated wire-frame assemblages objects, lang.: LISP	No boolean operations, no offsets, limited number of system defined features	Limited number of system defined features	Problems with "nonsense" parameters

Figure 5.1-2: Characteristics of a Subset of Advanced Modelling Systems.

The systems cover associative modelling techniques to some extent, and are described in greater detail in the following sections.

5.1.1 Language Based Approaches

ICAD is an example for a language based system. The language used is LISP like. Therefore, it can be also said, that it is a rule-based approach. The intention is to automate design decisions. A designer has to describe his decisions in terms of rules. The rules are written in a LISP like language.

A graphic browser is available for viewing the rules and a relational query language for retrieving parts from a library for use during the same phase of an actual design process (see Figure 2-3).

The design language is used to build up components and to define relative positioning of parts. There is also the possibility to create variants of product families. If the user wants to use these variants, he only has to specify the required parameter values and relative positions to combine and position them.

Products in the knowledge base are defined as hierarchical objects. This hierarchical organisation reflects the usual structure of a product in terms of levels of assemblies, sub-assemblies and components. The knowledge base contains generic object definitions as lists of attributes and rules. Information stored on a higher hierarchical level can be inherited by elements on a lower level. Initially, there is no data contained in the knowledge base: the user must completely define his own knowledge base for each type of product before he can use the system for design tasks.

The system does not include a geometric modeller. Geometric data is held in a seperate data base. Viewing transformations however can be applied for visualization on a graphic display (/AND-89/, /BRE-88/, /MAC-88/).

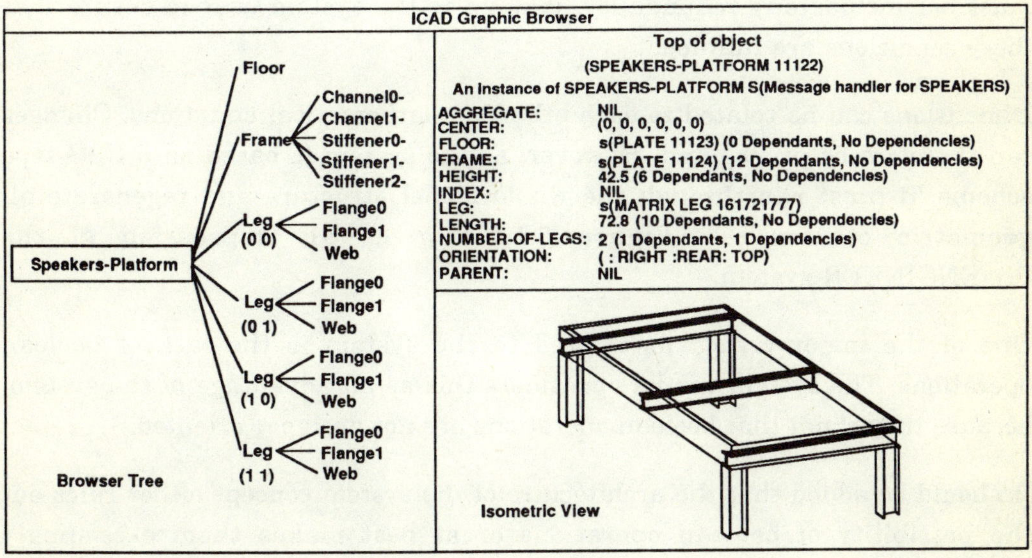

Figure 5.1.1-1: The ICAD System /AND-89/

5.1.2 Constructive Approaches

As examples for constructive approaches the Pro/ENGINEER system and the Cimplex System are described.

The Pro/ENGINEER system includes dimension-driven solid modelling techniques. Solid models are created by, and only by, sweep-type operations (and only by sweep-type operations). However, pre-defined form features can be used and additional, user defined form features can be stored in a feature library.

Modelling begins with the fully dimensioned definition of the cross-section of a part. In order to create a solid, the dimensions must be both sufficient and they must not include any redundancy. It is up to the system user to ensure that these conditions are fulfilled.

Dimensions can be related to each other by mathematical equations. Changes can therefore be propagated. However, as the system is based on a CSG-type scheme, it must run through the whole model structure and regenerate all geometric occurrences. Figure 5.1.2-1 gives an impression of the Pro/ENGINEER system.

One of the major criticisms related to the system is the lack of boolean operations. The system vendors considers this as an advantage of the system, because they claim that boolean operations are not designer oriented.

It should be added that the architecture of the system concept either rules out the possibility of boolean operations or at best makes them exceedingly inefficient

The vendors of the system further claim to support the whole of the product design process. However, as explained in Chapter 2, the design process involves being able to develop complex assemblies by starting with detailed components which are then integrated in a later design phase. Furthermore, it should be possible to insert parts, that were developed separately. Such technique requires the availability of boolean operations, which are not part of the Pro/ENGINEER system.

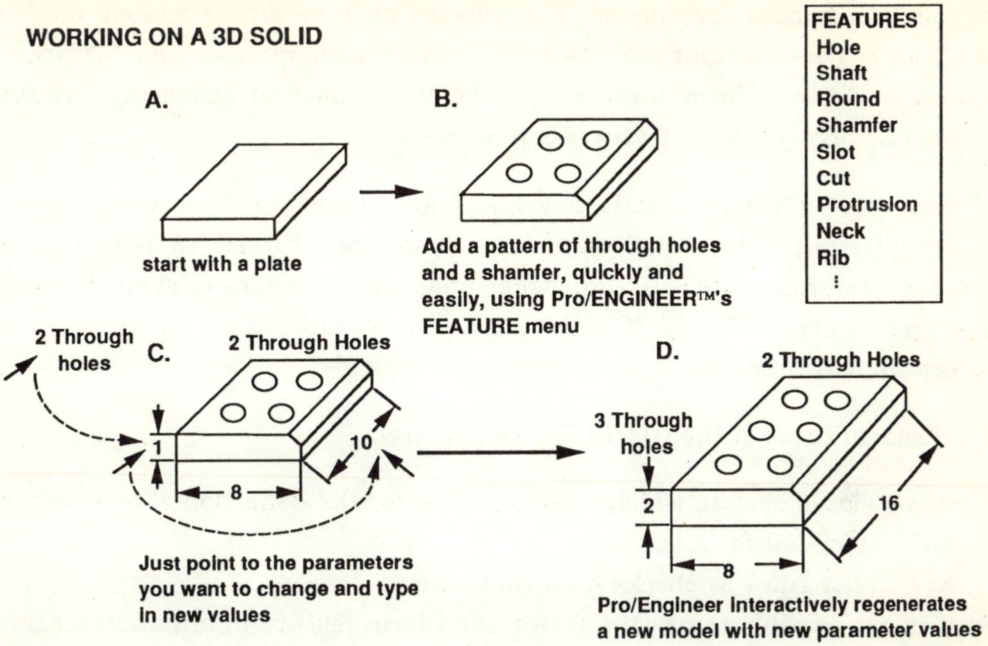

Figure 5.1.2-1: The Pro/ENGINEER System /FER-89/

CIMPLEX is a commercial CSG-modeller with a non-standard CSG concept. Form features appear as elements of the CSG-structure. It combines conventionel geometric design with the compositional "design by feature" approach. These form features are known as "implicit" or "hard" features and are part of the standard CIMPLEX product. These features are used in process planning and NC code generation. This approach restricts the use of form features to very specific pre-defined applications.

This library of form features can not be extended. The form features include a simple countersink, chamfered fillet, chamfer, pocket, cutout, boss and groove and are available for the entity classes, "hole", "edge", "recess", "protrusion" and "symmetric".

All machining features must be created as implicit form features. Explicit form features can also be created and are defined interactively by selecting entities from a boundary model. The selected entities can be named (e.g."side wall", etc). These names can then be used in user queries and for process planning. Explicit form features can be made part of other explicit form features i.e. explicit form features can be nested.

All classes of ANSI geometric tolerances are supported. Tolerances are not simply attributes but auxiliary data structures linked to features and geometric tolerances. ANSI Dimension and Tolerance (D&T) symbols can be displayed. Certain kinds of checks have been built-in to prevent invalid tolerances /CAM-88/.

The disadvantages of the CIMPLEX system are:

- it is a closed system which does not allow for the definition of new features and associated form features;
- no validity rules or checks for form features;
- it is not possible to use the feature and form feature information for other applications; the system cannot be extended.

5.1.3 Other Approaches

The recent version of the Intergraph system e.g. supports the use of associativities. Functional, mathematical, or logical relations can be defined in between objects of the model.

One application is the dimensioning of objects. Changing the object automatically updates the dimensions. The system is based on a uniform representation of geometry using NURBS: the associativities therefore also apply to B-Spline-surfaces /COH-88/.

5.2 The DTT-Modeller (Design by Technical Terms)

The DTT-Modeller has been developed as a prototype modeller to illustrate research results. DTT stands for Design by Technical Terms and represents a system which is able to demonstrate the principles of technical and especially technical-associative modelling.

Technical Modelling is defined as the modelling of engineering elements, i.e. products, assemblies, parts, form features as well as their attributes (position, orientation, dimensions, tolerances, materials, surface qualities, etc.). Technical-associative modelling is defined as the modelling of relationships between these engineering elements. These relationships often allow the parameter or attributes to be determined of engineering elements. Both, technical modelling and technical-associative modelling are required when building a set of modelling operations for the support of the preliminary design phases as described in Chapter 2. The development of these modelling operations involved augmenting the product model of the prototype system by a "technical shape" model. This support the modelling of

- products
- assemblies,
- parts,
- features and
- relationships between any of the above.

An equally important enhancement has also been introduced for the representation of:

- construction points,
- construction lines,
- construction faces,
- construction spaces.

The main technical modelling operations in the DTT-modeller are illustrated in Figure 5.2-1 and explained in greater detail below.

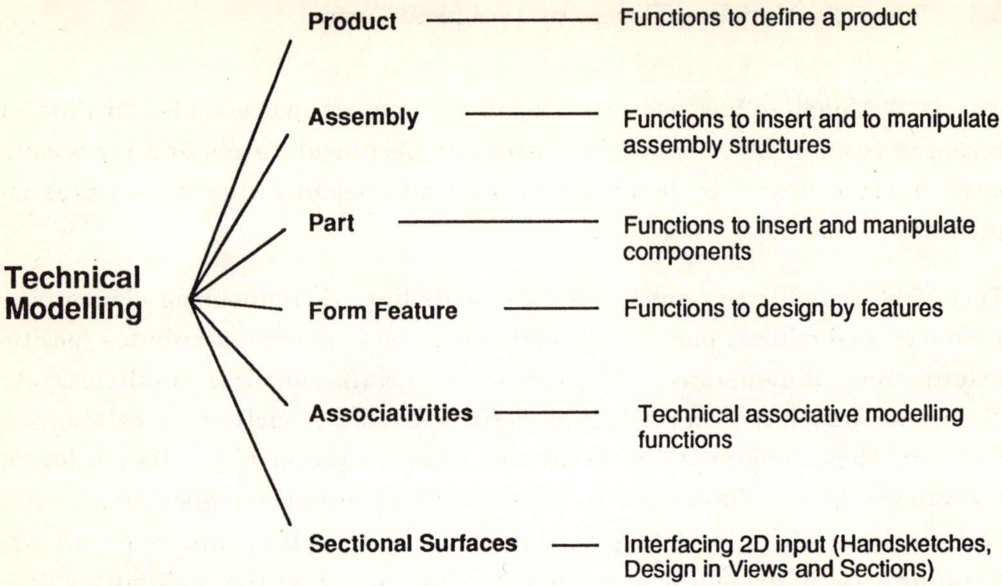

Figure 5.2-1: Operations for Technical Modelling in the DTT-Modeller

5.2.1 Sectional Surfaces

The solution of geometric problems often requires additional or sectional views of the relevant part. The result of the "handsketch" input system (HIS) consists of 2D-information. The effective use of a CAD-system requires the three dimensional modelling of engineering components. A sweep operator can be used to transform two dimensional data into full three dimensional solid models. This requires both, the 2-D "face" and a trajectory to fully define the swept solid. Many kinds of trajectories are possible. In the context of this project only translational (defined by a translation vector) and rotational trajectories (defined by a rotation axis and an angle) have been considered. This restriction is imposed by the modeller, which has only planar, cylindrical and conical surfaces. The definition of a sweep operation also requires certain conditions to be met, e.g., a face must not be moved along a vector, which is perpendicular to its normal vector; the result would not be a solid.

To understand the aspects of a technical sweep operator the technical details of the underlying modeller must be explained. The available modeller has more than just geometric elements. It also support associativities and organizational objects - such as form features, sub-assemblies or assemblies. The sweep operator must also transfer any engineering information from the input face to the created volume and store explicitly the semantics of the sweep operation itself.

The sweep operator transforms two dimensional geometric data into real solid modelling structures. Consequently, it will be used after programs which generate two dimensional data. Of course such programs should also insert any relevant engineering information. The output of the sweep operation is a complete boundary representation of a three dimensional object including all engineering information. The effective use of such models requires efficient 3D-operations. Figure 5.2.1-1 illustrate the kind of operation performed by a technical sweep.

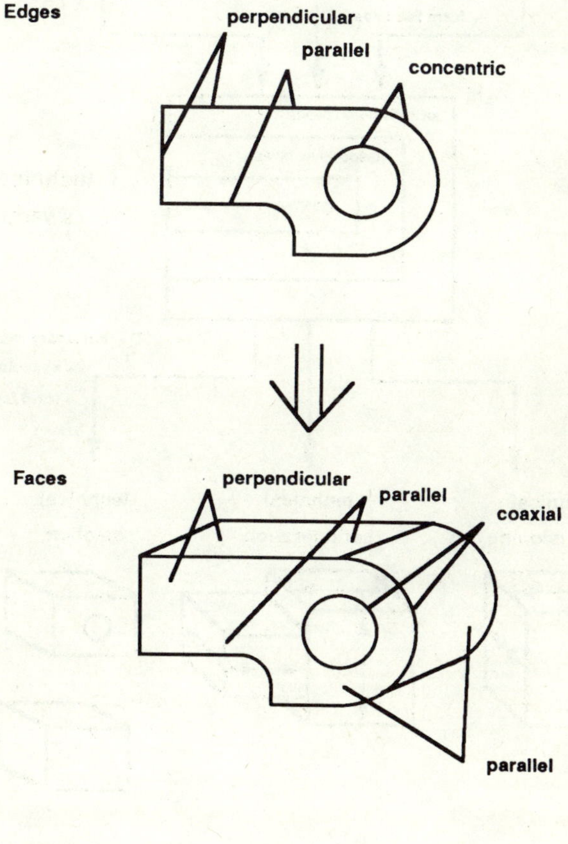

Figure 5.2.1-1: A Technical Sweep Operation

The technical sweep operator forms a link between 2D input and 3D representations in the DTT- Modeller: It therefore also forms the link between the "Handsketch" input system (HIS) and the DTT-Modeller itself. However, within the DTT-Modeller 2D structures can also created either interactively or by an application program. The 2D data can contain relationships or form features. Both of which are transformed into the corresponding 3D relationship or 3D features. This role of the technical sweep operation is illustrated in Figure 5.2.1-2.

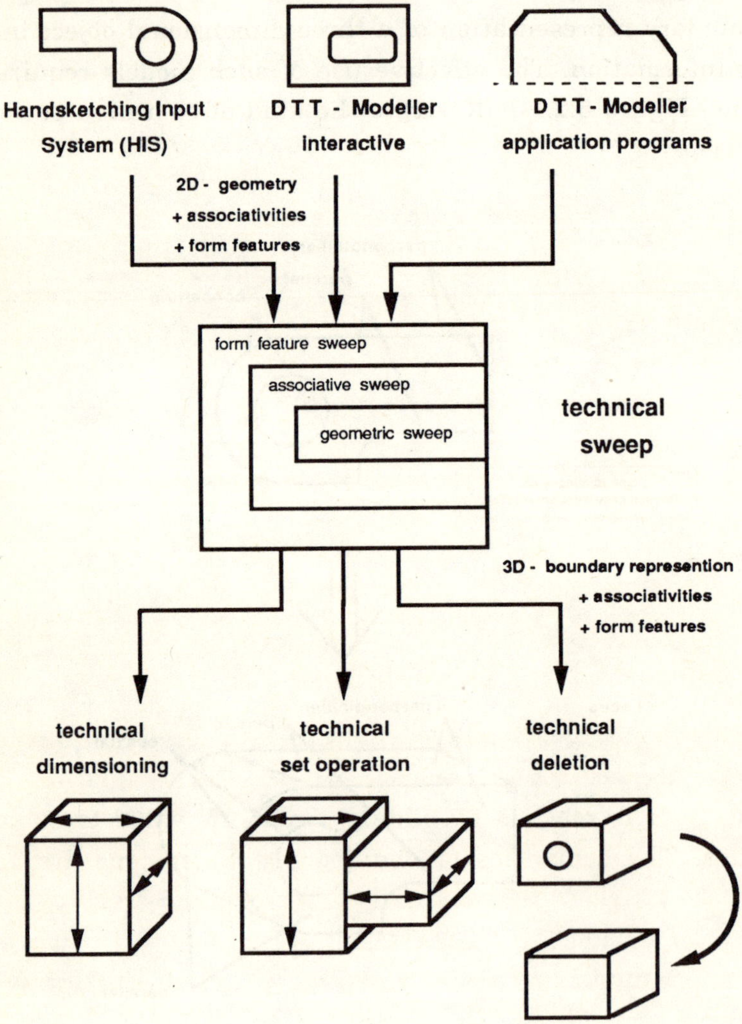

Figure 5.2.1-2: Use of the Technical Sweep Operation in the DTT-Modeller

5.2.2 Technical Associativities

Technical associativities are relevant both for the 3D dimensioning of a part and the use of technical terms for modelling assemblies. Possible associativities in a boundary representation were discussed in Chapter 3.2. The technical semantics will now be explored. Figure 5.2.2-1 describes the systematic use of geometric associativities for dimensioning purposes.

reference \ action		vertex	edge: line	edge: circle	face: plane	face: cylinder conical surface
vertex		distance of vertices —	dist. vertex - line —		dist. vertex - plane ----	
edge	line	dist. vertex - line —	perpendicular lines ---- distance of lines parallel angled lines angled		dist. of line and plane parallel angle line - plane angled	cylindrical coaxial conical coaxial
	circle				dist. of line and plane parallel	
face	plane	dist. vertex - plane —	dist. of line and plane parallel angle line - plane angled	dist. of line and plane parallel	distance of planes parallel angled planes angled	
	cylinder conical surface		cylindrical coaxial conical coaxial			

Legend : type of dimension / implicit geometric association

Figure 5.2.2-1: Use of Geometric Associativities for Dimensioning

The technical semantics of geometric associativities lead to a layered concept of constraints. The base layer is that of geometric associativities. The layer of metric constraints is reached by adding a numeric value to such geometric associativities. The semantics of engineering dimensions are represented exactly by these metric constraints. Hence, a technical modelling system must support the definition of constraints, take them into account during modification operations, and allow modelling via such constraints (e.g. as in dimension-driven modelling).

Technical restrictions form the third layer and involve placing constraints on constraints, e.g. by constraining the numeric parameter of a metric constraint. For example, the numeric values may be required to be equal and constant or to lie in a given internal of valid values. The semantics of engineering tolerances involve just such restrictions on metric constraints.

Functional constraints form a fourth layer by using mathematical equations to place constraints on constraints. Functional constraints will not be discussed further in this stage. Figure 5.2.2-2 gives an overview of the layers of constraints.

The constraints relevant to technical-associative modelling can be derived directly from the user-input, provided a fixed list of technical terms is used to communicate with the advanced CAD-system. A relevant list of technical terms is:

- adjacent
- angular
- centered
- concentric
- flush
- symmetric
- aligned
- axially parallel
- coaxial
- congruent
- parallel

Layers of Constraints		Technical Semantic	Examples	
4	Functional Constraints	Physical Equations	calculated distance	d = f (x)
3	Technical Restrictions	Technical Tolerances	toleranced distance	$d^{+0,2}_{-0,4}$
2	Metric Constraints	Technical Dimensions	distance	d
1	Geometric Association	Implicit constraints	parallel	//

Figure 5.2.2-2: Layers of Constraints

The prototype implementation illustrates the following features:

1. Technical-associative terms are used to generate the parameters for modification operations. They can be used to provide positions, orientations and dimensions.

2. Technical-associative terms are used to insert a part into an assembly-structure.

3. Technical-associative terms create relationship on the various entities of an assembly. These relationships are stored explicitly in the advanced CAD model and are therefore available for subsequent modelling operations.

The example in Figure 5.2.2-3 illustrates the effect of technical-associative modelling when using the term "aligned".

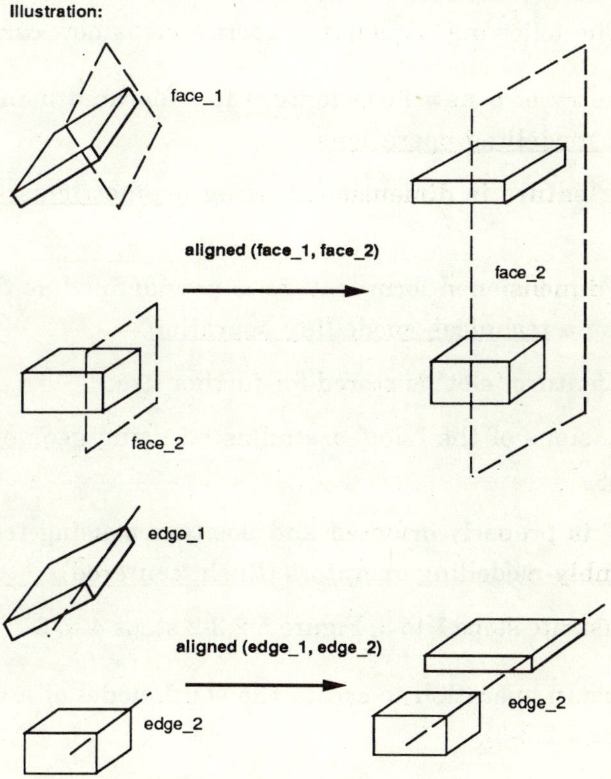

5.2.2-3: Effects of the Technical Term "aligned"

When modelling with technical terms their context sensitive semantics must be taken into account: e.g. the term "aligned" when applied to faces has a different effect than when it is applied to edges. Furthermore some degrees of freedom remain unconstrained: These must therefore either be chosen arbitrarily, or determined from the current context.

5.2.3 Form Features and Parts

Associative methods can be used in addition to conventional methods when modelling single parts and form features. For example form features can be described using associativities and then modified in terms of these.

The following example illustrates this approach.

The starting position consists of a single part whose dimensions are given (see Figure 5.2.3-1). The following sequence of operations is now carried out:

1. The geometry of a new form feature is generated using conventional geometric modelling operations.
2. The form feature is dimensioned using geometric-associative operations.
3. The fully dimensioned form feature is now defined as the form feature "slot" using a technical- modelling operation.
4. The form feature "slot" is stored for further use
5. The dimensions of the "slot" are adjusted using geometric-associative operations.
6. The "slot" is properly oriented and positioned using technical-associative assembly-modelling operators (flush, centered)

Figure 5.2.3-1 illustrate steps 1 to 3, Figure 5.2.3-2 steps 4 to 6 .

7. Use a boolean operation to create the solid model of a cube with a slot (see Figure 5.2.3-3).

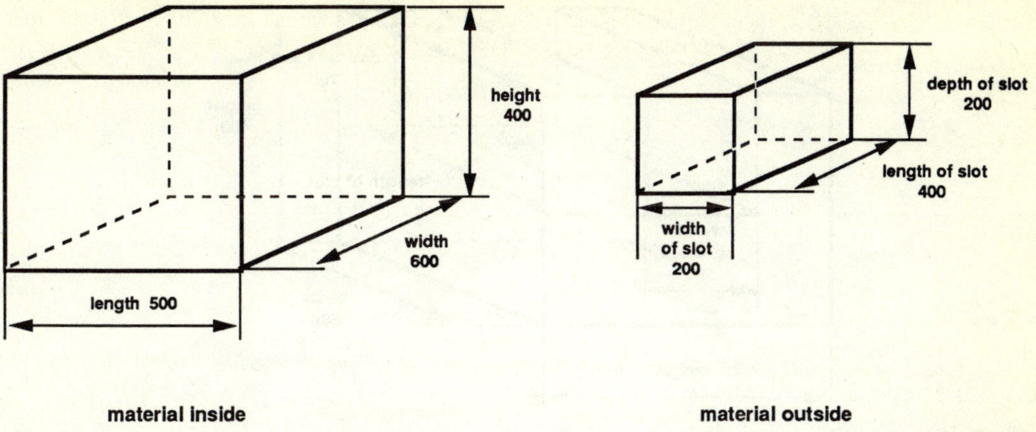

Figure 5.2.3-1: A Cube and Slot

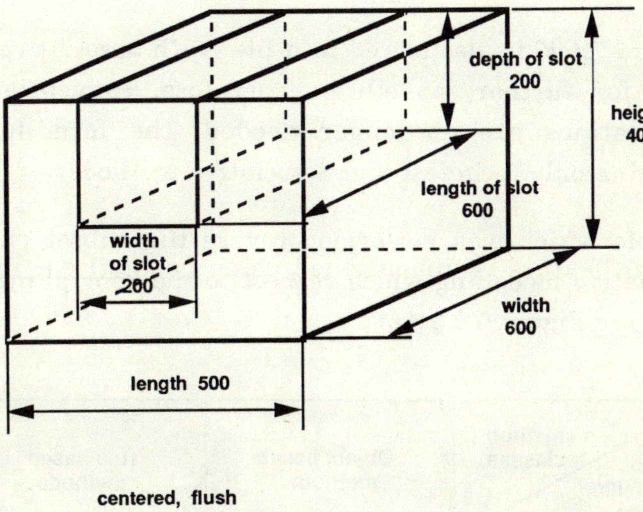

Figure 5.2.3-2: The form feature "slot" is properly Dimensioned, Oriented and Positioned with respect to the Cube

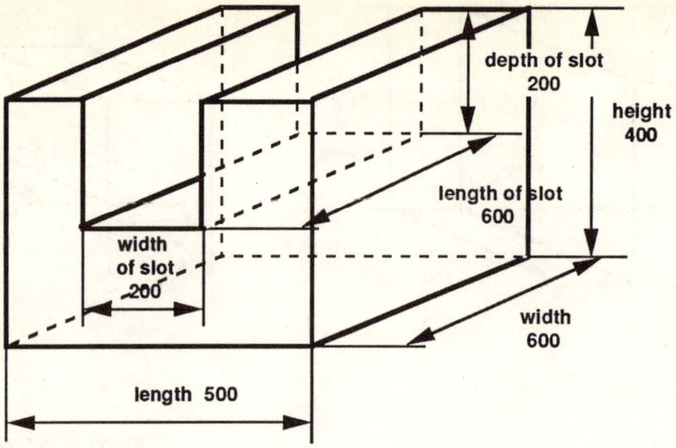

Figure 5.2.3-3: *Result of Boolean Operation*

The form feature "slot" is also stored in a file (with associative relationships) and available for further modelling. Therefore, conventional geometric modelling operations are no longer needed: the form feature can be manipulated using only technical and associative methods.

The design cycle which was performed covers the subset of technical and technical-associative modelling which respect to the general modelling matrix in Figure 3.2.1 (see Figure 5.2.3-4).

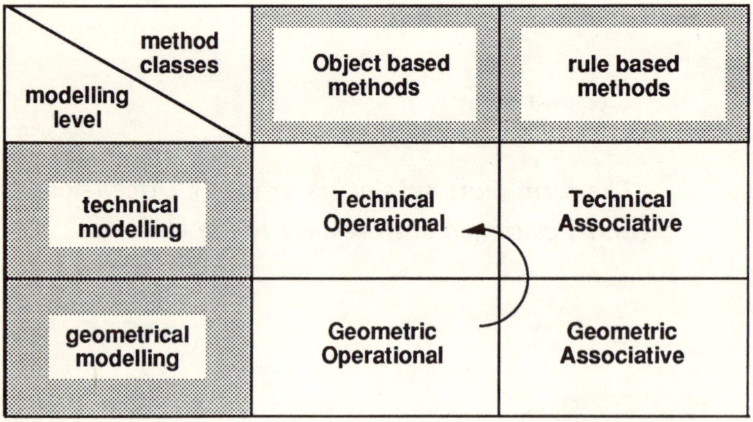

Figure 5.2.3-4: *Covered Subset of the General Modelling Matrix*

5.2.4 Assemblies and Products

When an associative term is used its semantics are also analyzed with respect to the modelling of assemblies. The effect of an associative term depends on the types of the parts it is applied to. This phenomenon occurs whenever existing entities are integrated into the product structure. The product structure is itself represented by a data structure. A high-level outline of the data-structure is shown in Figure 5.2.4-1.

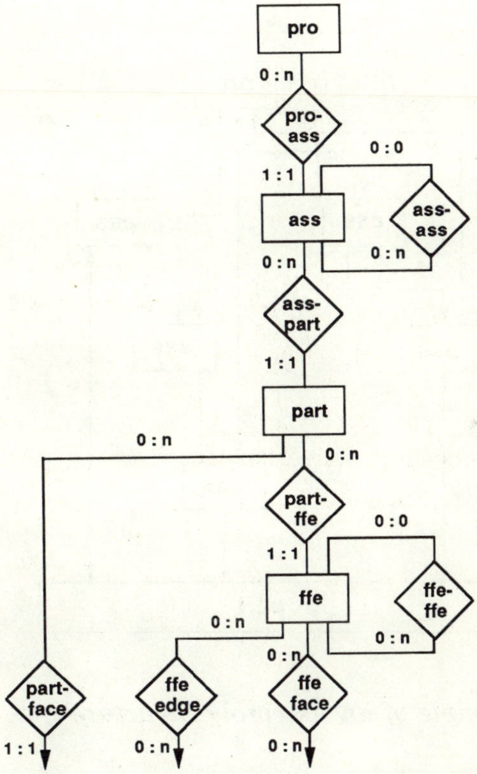

Figure 5.2.4-1: Product Structure

The product "pro" is the top-level entity. It consists of top-level assemblies "ass" which can themselves consist of a hierarchy of sub-assemblies (Relationship "ass-ass"). A part "part" belongs to exactly one assembly, but an assembly or a subassembly may contain one or more parts or sub- assemblies. A part is described by several faces "face" (a boundary representation scheme is needed). At the same time a part may also be described in terms of form features "ffe". Again, there are top-level features. The structure of feature-subfeatures relationships is the same as that of product-assembly relationships and involves a hierarchy of relationships "ffe-ffe". A feature may contain one or more faces or sub features.

The following figures will provide examples of actual product data structures. Figure 5.2.4-2 describes an assembly data structure, whereas Figure 5.2.4-3 describes a complex feature data structure.

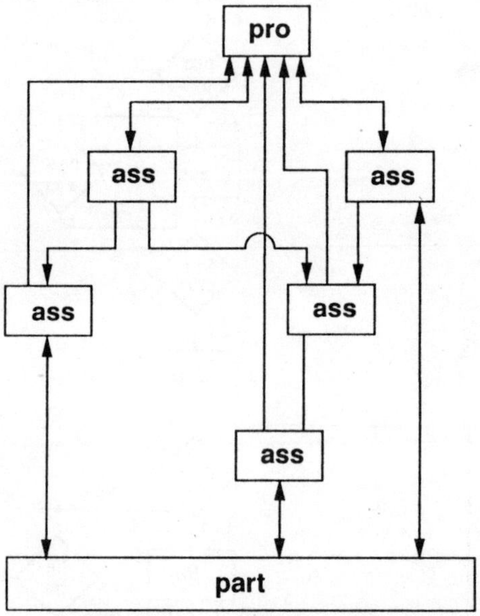

Figure 5.2.4-2: Example of an Assembly Structure

Both data structures require a top-level entity which ensures the uniqueness of the description.

The use of associative terms requires knowledge of the product data structure. At the same time, it allows information to be added to the product structure. The following matrix shows the parts of the product structure which can be generated by associative terms depending on the type of action and reference entity (Figure 4.2.4-4).

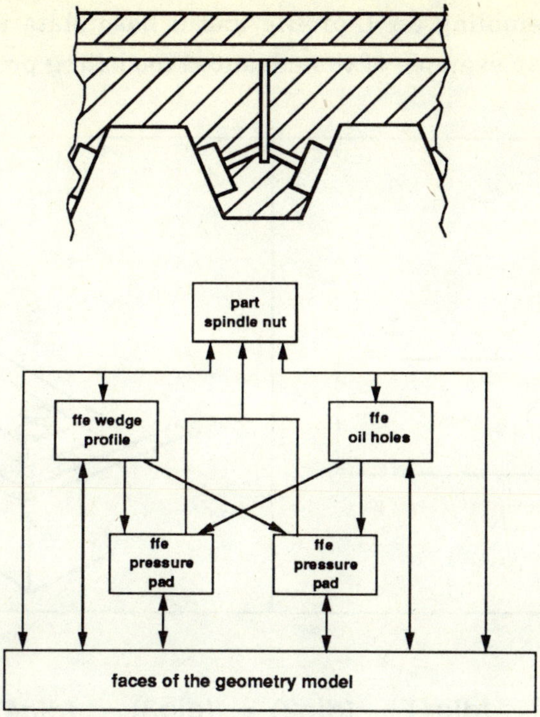

Figure 5.2.4-3: Example of a Feature Structure

Reference \ Action	Form Feature	Part	Assembly	Product
Form Feature	Feature Subfeature	Assembly Part Feature	Assembly Subassembly Part Feature	Product Assembly Subassembly Part Feature
Part	Part Feature	Assembly	Assembly Subassembly	Product Assembly Subassembly Part
Assembly	/	Assembly Part	Assembly Subassembly	Product Assembly Subassembly
Product	/	/	Product Assembly	/

Figure 5.2.4-4: Subsets of the Product Data Structure, which can be generated through the Use of Associative Terms

An example of assembling an L-profile and a base-plate is shown in figure 5.2.4-5 using an early example of the advanced modelling prototype system.

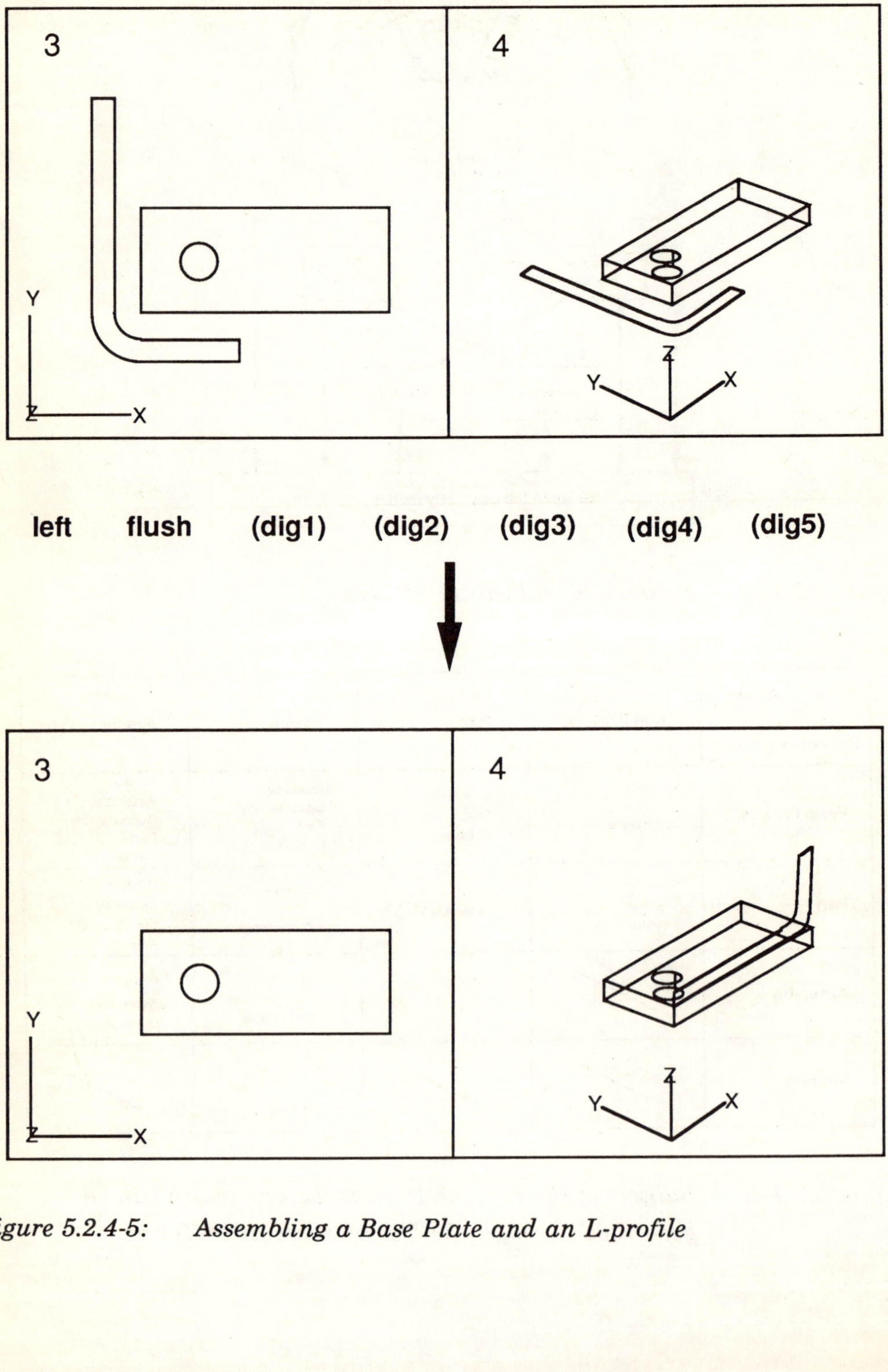

Figure 5.2.4-5: Assembling a Base Plate and an L-profile

5.3 Command Language for Design

The overlapping question on advanced modeling is: How does the designer communicate with the CAD system? Two major approaches are being persued simultaneously in the project as shown in Figure 5.3-1

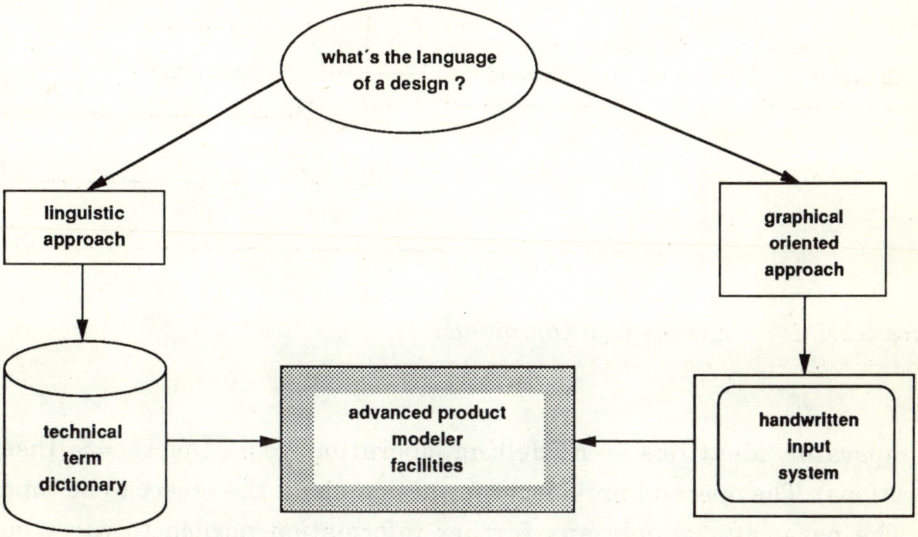

Figure 5.3-1: Approaches to Advanced Modelling

The linguistic approach comprises the set of all CAD system command languages. The graphical approach includes all graphical communication techniques. These are the two main approaches for CAD system communication. Any effort to improve such communication must therefore involve improving one or both of these approaches.

A higher level of communication on the graphical side can be obtained by allowing "handsketch" input into the CAD system which not only support the input of sequences of points, but also recognizes graphical elements and associativities between them. On the linguistic side the input of natural language can be considered a suitable goal, but requires prior classification of the semantics of technical terms.

5.3.1 Command Syntax

Commands can be described by syntax diagrams. In general they have the following form (Figure 5.3.1-1).

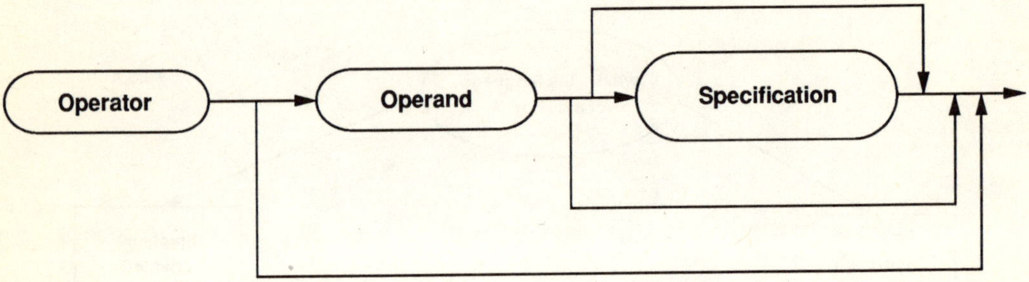

Figure 5.3.1-1: Syntax of a command

The operator identifies a modelling operation (e.g. Insert => Insertion Operations). The operand provides information about the object to be "operated on". The parameter supply any further information needed to carry out the operation.

Using this principle, four classes of terms were found to be necessary.

Class 1: Operators

Definition: Terms or combination of terms, which identify a specific modelling operation.

Examples: Insert
Generate
Build => Operation for inserting objects
Join
Unify => Operation for obtaining the union of two objects
Add => Operation for inserting an object and for combining it with an existing object

Class 2: Operands

Definition: Terms or combination of terms, which can be stored in the model as occurrences of the product model specification.

Examples:
straight
line
circle
cylindrical face
solid block
chamfer
dowel

Class 3: Attributes (a kind of parameter)

Definition: Command words which are necessary to specify parameters of operands

Examples:
Length =
Position =

Class 4: Associations (a kind of parameter)

Definition: Command words which can be evaluated to generate arguments for modelling operations

Examples:
parallel
coaxial
flush
fitted in

An overview of these classes of command words used in the project is given in Figure 5.3.1-2.

Components	Operators	Operands	Attributes	Associativities
Definition	Terms which select a modelling method in principle	Terms which can be stored as occurences of the model scheme	Terms which specify parameters of operands	Terms which can be evaluated to generate attributes
Instances	**o insertion** - insert - generate - sweep **o interrogation** - get **o modification** - modify - join, unify - intersect - subtract **o deletion** - delete	**o geometric information** - vertices - edges - faces - volumes **o technical information** - form feature - parts - assemblies	**o geometric attributes** - distances (lenght, width, height, thickness) - angles - radii diameters **o technical attributes** - surface finishes - tolerances - material	**o geometric associations** - parallel - angular - concentric - symmetric **o technical associations** - coaxial - flush - imposed - aligned - fitted in

Figure 5.3.1-2: Language Elements

5.3.2 Technical Modelling Language

The communication interface of the prototype modeller currently supports a subset of the above commands. This subset is related to the set of operands available in the current modeller. The operands are:

- product
- assembly
- part
- form feature
- associativity

The following is a list of the operations and their parameters needed to perform a specific modelling function :

```
product (pro)
     create <pro_data>
          A new product entity is generated.
     define <ass_ident> <pro_data>
          A new product entity is generated.
          A set of assemblies are grouped to form a product.
     undefine <pro_ident>
          All references of the product entity
          to its related entities are deleted.
     undefine and delete <pro_ident>
          A product entity is deleted.
          Entities referenced by the product are not deleted.
     modify <pro_ident> <pro_data>
          Data of a product entity are changed.
     delete <pro_ident>
          A product entity together with all entities
          belonging to it is deleted.
     translate <pro_ident> <vector>
     rotate <pro_ident> <point> <axis> <angle_degree>
     copy <pro_ident>
          A copy of an entity together with all its
          sub-entities is created.
     store <pro_ident> <file_data>
     load <file_choice>

assembly (ass)
     create <ass_data>
          A new assembly entity is generated.
     insert <ass_choice>  <generic_ident(pro,ass)>
          From a choice of assemblies one is selected.
          The chosen assembly and all its entities are
          inserted into the current data structure as well
          as related with an existing product or assembly.
     define <generic_ident(ass,part)> <ass_data>
          A new assembly entity is generated.
          A set of entities are grouped to form an assembly.
     undefine <ass_ident>
          All references of the assembly entity
          to its related entities are deleted.
     undefine and delete <ass_ident>
          An assembly entity is deleted from a treelike assembly
          structure.Entities referenced by the deleted assembly
          are not be deleted.They now are referenced the next
          higher entity in the hierarchy.
     modify <ass_ident> <ass_data>
```

Data of an assembly entity are modified.

delete <ass_ident>

An assembly entity together with all its entities is deleted.

add <ass_ident> <ass_data>

A new assembly entity is added to an existing assembly data structure .

split and add <ass_ident> <ass_ident> <ass_data>

A new assembly entity is added between two assembly entities in an assembly structure.

translate <ass_ident> <vector>

rotate <ass_ident> <point> <axis> <angle_degree>

scale

copy

A copy of an entity together with all its entities is created.

store <ass_ident> <file_data>

load <file_choice>

part (part)

create <part_data>

A new part entity is generated.

insert <part_choice> <ass_ident>

From a choice of parts one is selected. The chosen part and all its entities are then inserted into the current data structure as well as related with an existing assembly-structure.

define <generic_ident(face,ffe)> <part_data>

A new part entity is generated. A set of entities are grouped to form the part.

undefine <part_ident>

All references of the part entity to its related entities are deleted.

undefine and delete <part_ident>

A part entity is deleted. Referenced entities are not deleted.

modify <part_ident> <part_data>

Data of a part entity are changed.

delete <part_ident>

A part entity together with all its entities is deleted.

translate <part_ident> <vector>

rotate <part_ident> <point> <axis> <angle_degree>

copy <part_ident>

A copy of an entity together with all its entities is created.

create_hull <part_ident>

A hull is created around the identified part.

```
        The new object has a partlike character.
        All operations for parts are also available
        the manipulation of hull spaces.
        However, the constraint of being a hull must still be
        fulfilled afterwards.
    create_complement <part_ident>
        Changes the material direction.
    unify, intersect, subtract <part_ident> <part_ident>
        More than one part can be modified by these boolean
        operations.
    store <part_ident> <file_data>
    load <file_choice>

form feature entity (ffe)
    create <ffe_data>
        A new form feature entity is generated.
    insert <ffe_choice> <generic_ident (part,ffe)>
    define <generic_ident(face,ble)> <ffe_data>
        A new form feature entity is generated.
        A set of entities are grouped to form the feature.
    undefine <ffe_ident>
        All references of the form feature
        to its entities are deleted.
    undefine and delete <ffe_ident>
        An form feature entity is deleted from a treelike
        form feature structure. Entities referenced by the
        deleted form feature are not be deleted.They now are
        referenced the next higher entity in the hierarchy.
    modify <ffe_ident> <ffe_data>
        Data of a form feature entity is modified.
    delete <ffe_ident>
        A form feature entity together with its entities
        is deleted.
    add <ffe_ident> <ffe_data>
        A new feature entity is added to an existing feature
        structure.
    split and add <ffe_ident> <ffe_ident> <ffe_data>
        A new feature entity is added between two feature
        entities in a existing feature structure.
    translate <ffe_ident> <vector>
    rotate <ffe_ident> <point> <axis> <angle_degree>
    scale

    copy
        A copy of an entity with all its
        entities is created.
    create_hull <ffe_ident>
        A hull is created around the identified form feature.
        The new object has a form featurelike character.
```

```
          All operations for form features are also available
          the manipulation of hull spaces.
          However, the constraint of being a hull must still be
          fulfilled afterwards.
     store <ffe_ident> <file_data>
     load <file_choice>

technical associativity entity (tae)
     insert <tae_choice>  <tae_ident>
          The chosen associativity is
          inserted into the current data structure.
     undefine and delete <tae_ident> <tae_data>
          An associativity entity is deleted.
          All entities referenced by the associativity/dimension
          are not deleted.
     modify <tae_ident> <tae_data>
          Data of an associativity entity are
          modified.
```

5.3.3 Technical Associativities

In future, the set of associativities will have to be augmented and customized to suit the special user needs. Therefore, a fixed set of predefined terms is not suited to technical associative modelling. Another approach is therefore pursued in the project, namely the definition of a language for defining the meaning of associative terms /GAR-86/.

This language is specified below. Definitions of technical-associative terms have been made.

Formal Specification of Technical-Associative Terms

Technical associativities involve complex data manipulations. It is therefore necessary to describe associativities in a formal way. The formal definition also makes the computer implementation easier. The result is the following formal language which is defined using an extended Backus-Naur-notation. We use the following production and symbols where the capital letters represent production of the formal language:

A: definition of associativity A

(A | B): use of production A or B alternatively (exclusively)

(A)nm: production A can be repeated from n to m times where n>=0, m>=0.
If m is a "*" then A can be repeated infinitely many times.

All other symbols are elements of the formal language. In this case every associativity can be defined by a sequence of expressions:

. associativity :== (expression | [expression] ((^ | v)[expression])1*)

The associativity itself consists of its name and a list of required variables:

associativity: (name)1* ((list_of_variables))0*

name: (a | b | c... | z)(a | b | c... | z | 0 | 1... | 9 | _ |)0*

list_of_variables: variable(,variable)0*

To make the variables more understandable its names consist of its type followed by a number. This allows debugging during runtime too.

expression: (condition((^ | v)condition)0*)01 => action

condition: (A=B | A≠B | A<B | A<=B | A>B | A>=B | C)

where A and B are of the same type.

A,B: (function | variable | constant)

C: (procedure | associativity)

action: (procedure | associativity | assignment)11

function: atom (list_of_variables)

procedure: atom ((list_of_variables)01

assignment: A=B

Now the expressions contain the meaning of the associativity using programs (called atoms) or other associativities. (Note: recursive associativities are not supported.)

Every associativity has an implicit parameter of success which can be used in the calling "associativity". So very complex structures can be expressed in a very intuitive way. If a condition is fulfilled the related actions will be performed and the success parameter is set to "true". If no condition is fulfilled or the atoms detect an error the success parameter is set to "false".

The expressive power of this language will be illustrated in the following examples. The main advantage is the modular definition of associativities. For this reason many associativities can be built from a limited set of built-in functions like "type", "parallel" etc. New associativities can be added without the need for new atoms. However, additional atoms for creating and deleting elements of the CAD system data structure. (Note: The better the data structure of the CAD system the better associativities can be implemented)

```
adjacent

    adjacent(OFE1,OFE2,BLE1,BLE2)  :==
       type(OFE1)  = PLANE
       ^type(OFE2)  = PLANE
    => plain_parallel(OFE1,OFE2,0)
       parallel(BLE1,BLE2,0)

aligned

    aligned(OBJ1,OBJ2)  :==
    [  type(OBJ1)  = edge
       ^type(OBJ2)  = edge
       ^OBJ1 =/ OBJ2
    => line(OBJ1)  = line(OBJ2)       ]
    [  type(OBJ1)  = face
       ^type(OBJ2)  = face
       ^OBJ1 =/ OBJ2
    => surface(OBJ1)  = surface(OBJ2)     ]

aligned to

    aligned to (OBJ1,OBJ2)  :==
       => aligned(OBJ1,OBJ2)

angular

    angular(OBJ1,OBJ2,ANGLE)  :==
    [  type(OBJ1)  = ROTATION_OBJ
       ^type(OBJ2)  = ROTATION_OBJ
    => LINE1 = rotation_axis(OBJ1)
       LINE2 = rotation_axis(OBJ2)
       insert_angle(LINE1,LINE2,ANGLE) ]
```

```
    [  type(OBJ1) = PLANE
       ^type(OBJ2) = PLANE
    => SURFACE1 = surface(OBJ1)
       SURFACE2 = surface(OBJ2)
       insert_angle(SURFACE1,SURFACE2,ANGLE) ]
    [  type(OBJ1) = LINE
       ^type(OBJ2) = LINE
    => LINE1 = line(OBJ1)
       LINE2 = line(OBJ2)
       insert_angle(LINE1,LINE2,ANGLE) ]
```

axially parallel

```
    axially parallel(OBJ1,OBJ2,DIST) :==
       type(OBJ1) = ROTATION_OBJ
       ^type(OBJ2) = ROTATION_OBJ
    => LINE1 = rotation_axis(OBJ1)
       LINE2 = rotation_axis(OBJ2)
       parallel(LINE1,LINE2,DIST)
```

centered

```
    centered(OBJ1,OBJ2) :==
    [  type(OBJ1) = BLE
       ^type(OBJ2) = BLE
       OBJ1 =/ OBJ2
    => center(OBJ1) = center(OBJ2) ]
    [  type(OBJ1) = OFE
       ^type(OBJ2) = OFE
       OBJ1 =/ OBJ2
    => center(OBJ1) = center(OBJ2) ]
```

coaxial

```
    coaxial(OBJ1,OBJ2) :==
    [  type(OBJ1) = ROTATION_OBJ
       ^type(OBJ2) = ROTATION_OBJ
    => GER2 = rotation_axis(OBJ2)
       rotation_axis(OBJ1) = GER2 ]
    [  type(OBJ1) = ROTATION_OBJ
       ^type(OBJ2) = GER
    => GER(OBJ2) = rotation_axis(OBJ1) ]
```

coaxial to

```
    coaxial to(OBJ1,OBJ2) :==
    => coaxial(OBJ1,OBJ2)
```

concentric

```
concentric(OBJ1,OBJ2) :==
[ type(OBJ1) = CIRCLE
  ^type(OBJ2) = CIRCLE
=> NVE(OBJ1) = NVE(OBJ2)  ]
[ type(OBJ1) = ROTATION_OBJ
  ^type(OBJ2) = ROTATION_OBJ
=> coaxial(OBJ1,OBJ2)            ]
[ type(OBJ1) = CIRCLE
  ^type(OBJ2) = ROTATION_OBJ
=> line(nve(OBJ1)) = rotation_axis(OBJ2) ]
[ type(OBJ1) = ROTATION_OBJ
  ^type(OBJ2) = CIRCLE
=> rotation_axis(OBJ1) = line(nve(OBJ2)) ]
```

congruent

```
congruent(FACE1,FACE2) :==
=> FACE1 = FACE2
```

flush

```
flush(FACE1,FACE2,EDGE1,EDGE2) :==
   type(FACE1) = PLANE
   ^type(FACE2) = PLANE
   FACE1 =/ FACE2
=> surface(FACE1) = surface(FACE2)
   partial_equal(EDGE1,EDGE2)
```

perpendicular

```
perpendicular(OBJ1,OBJ2) :==
=> angular(OBJ1,OBJ2,90)
```

plain parallel

```
plain parallel(OBJ1,OBJ2,DIST) :==
   type(OBJ1) = PLANE
   ^type(OBJ2) = PLANE
   OBJ1 =/ OBJ2
=> SURFACE1 = surface(OBJ1)
   SURFACE2 = surface(OBJ2)
   parallel(SURFACE1,SURFACE2,DIST)
```

```
symmetric

    symmetric(OBJ1,OBJ2) :==
       type(OBJ1) = type(OBJ2)
    => SURFACE1 = symmetry_plane(OBJ1,OBJ2)
       OBJ1 = reflection(OBJ2,SURFACE1)


list of atoms used

     type(OBJ1)            :    obtain the type of an object
                                (e.g. straight line)
     rotation_axis(OBJ1)   :    obtain the axis of symmetrie of an
                                object
     parallel(x,x,DIST)    :    insert the associativity
                                "parallel with characteristic
                                distance DIST"
     surface(FACE1)        :    identify a surface
     partial_equal(x,x)    :    make the given objects partial
                                equal
     nve(OBJ1)             :    get the normal surface vector of
                                an object
     line(OBJ1)            :    get the edge geometry
     insert_angle(x,y,
                  angle)   :    insert the given angle ANGLE
                                between the given objects
     center(OBJ)           :    get the center point of
                                object
```

The possibility of using HDSL /SCH–89/ or EXPRESS /ISO-89/ for the purpose of specifying associativities was considered.

The above language for the definition of technical terms was compared to EXPRESS. This was done by taking a common known language like Pascal and comparing the different language components with each other (see figure 5.3.3-1).

Looking at the Figure it is clear, that EXPRESS is used to describe type declarations of programming languages. However, in comparison to Pascal, EXPRESS has more powerfull means of specifying user-defined types. A record structure e.g. containing a case-sensitive definition of record-components is described linked by the keyword "subtype" (and "supertype" the other way round.)

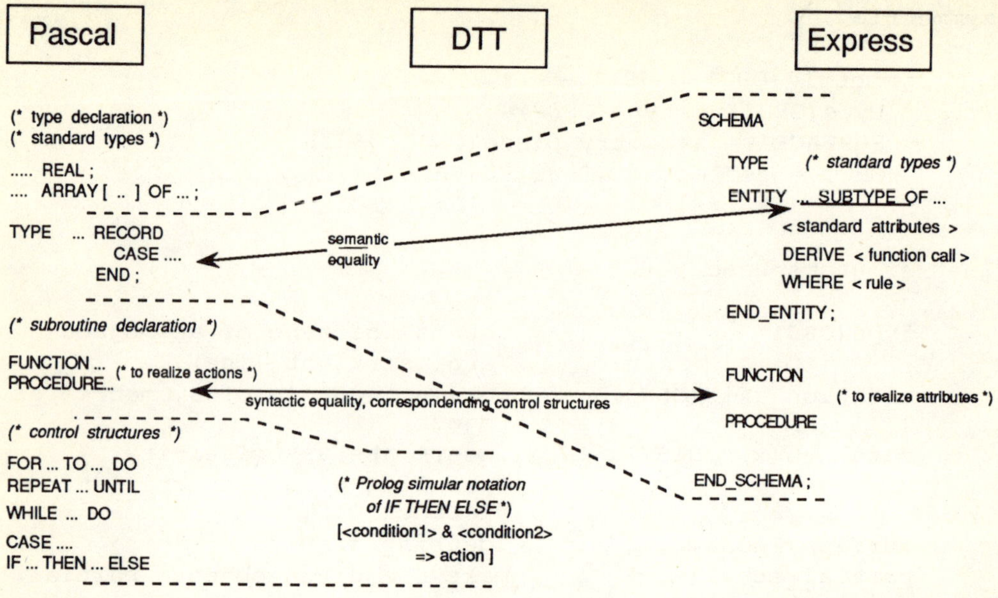

Figure 5.3.3-1: Comparison of Language Components

On the other side the DTT language was defined as a subset of Prolog and therefore corresponding to the control structures of Pascal. However, where Pascal has a variety of control structures, the DTT language contains only the Prolog similar notation of a nested if-then-else structure.

All Pascal control structures are available in EXPRESS. It is therefore possible, in principal, to take EXPRESS as language for the formal specification of technical terms. However, it may be necessary to extend EXPRESS by introducing an entry point like element (the "begin" of a procedural block in Pascal) to encompass all features of the above technical term language. HDSL is similar to EXPRESS. The same arguments therefore also apply to HDSL. For these reasons neither HDSL nor EXPRESS were chosen for the definition language.

The main effort of the later phase of the project was placed on realizing functions like "make geometric surfaces of two faces parallel, parallel with distance 0, identical, etc." These functions are the basic functions of an associative modeller and are called the "geometric-associative" functions. The interface specification for the complete set of functions used in the project is given in Chapter 7. An example (the function "aligned") is given in Figure 5.3.3-2.

Illustration:

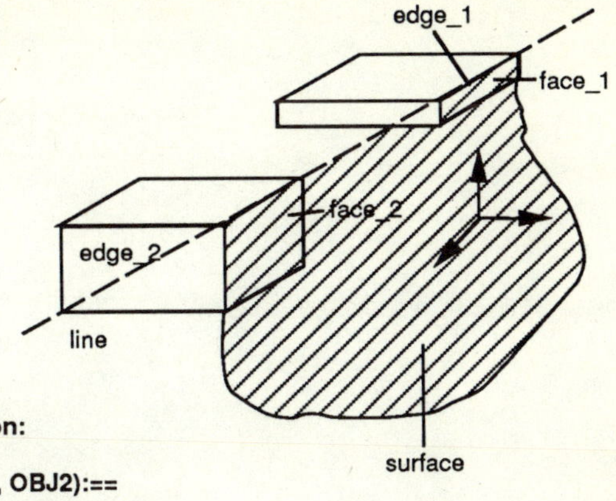

Formal Term Definition:

aligned (OBJ_1, OBJ2):==

* First case: alignment of edges

[type (OBJ_1 = edge

^ type (OBJ_2 = edge

^ OBJ_1 =/ OBJ_2

* make the underlying geometric structure identical
* taking care of given restrictions by moving the whole object

=> line (OBJ_1) = line (OBJ_2)]

* Second case: alignment of surfaces

[type (OBJ_1 = face

^ type (OBJ_2 = face

^ OBJ_1 =/ OBJ_2

* make the geometric faces of both surfaces identical

=> surface (OBJ_1) = surface (OBJ_2)]

aligned to (OBJ_1, OBJ_2) :== aligned (OBJ_1, OBJ_2)

* this is a synonym for `aligned (OBJ_1, OBJ_2)´

Legend: OBJ = object (variable of unknown type)
type ( ) = function to get the type of an object
line ( ) = function to get the geometric line of an edge
surface ( ) = function to get the surface to a given face
=/ = not equal
:== = term definition
( ) = variable
[ ] = expression
^ = conjunction
=> = action

Figure 5.3.3-2: Example the Function "aligned" described in the Definition Language

6. Design by Features

Stefan Rude and Mike Pratt

Design by features is an example of a conventional engineering design method. Form features are a large and complex area of study. Pratt /PRA-87/ showed that features are used, in design, process planning, NC data generation, dimensioning and tolerancing, inspection, etc. Although form features are often discussed, there is no common definition. Most authors say: "A form feature is a region of interest on the surface of a part" (see Figure 6.1).

Form Feature

A region of interest on the surface of a part

Different Views

Design

Analysis

Manufacture

Quality Assurance

Figure 6-1: Definition and Views of Form Features

The advantage of such a definition of form features is that it leaves form features the freedom to not only consist of collections of faces but also, for example, include edge modifiers (e.g "bevelled"), surface finish, tolerances, feature specific dimensions, or even feature specific associativities to related objects, (e.g. functional interfaces between parts or features).

The latest definition given by Prof. Pratt /PRA-88/ (see Figure 6-2) deals with the issue, that form features can be defined independently of solid modelling technology (e.g. B-Rep, CSG etc.).

What Is a Form Feature?

A related set of elements of
a product model, conforming to
characteristic rules allowing
its recognition and classifi-
cation, and which, regarded as
an independent entity, has some
function during the life cycle
of a product

Figure 6-2: What is a Form Feature ?

Form features created by the designer do not necessarily correspond to features required for other operations: some kind of automatic feature recognition (or at least feature translation) will be needed in future integrated systems.

There are two possibilities for handling features:

- offer a set of form features during product design, for example using a number of form features.
- complete the geometric design first and only then analyse it in order to extract the relevant features (ex-post analysis).

Two methods are available for ex-post analysis:

- an interactive manual specification of form features.
- an automatic analysis of a geometrically completely described part (see Figure 6-3).

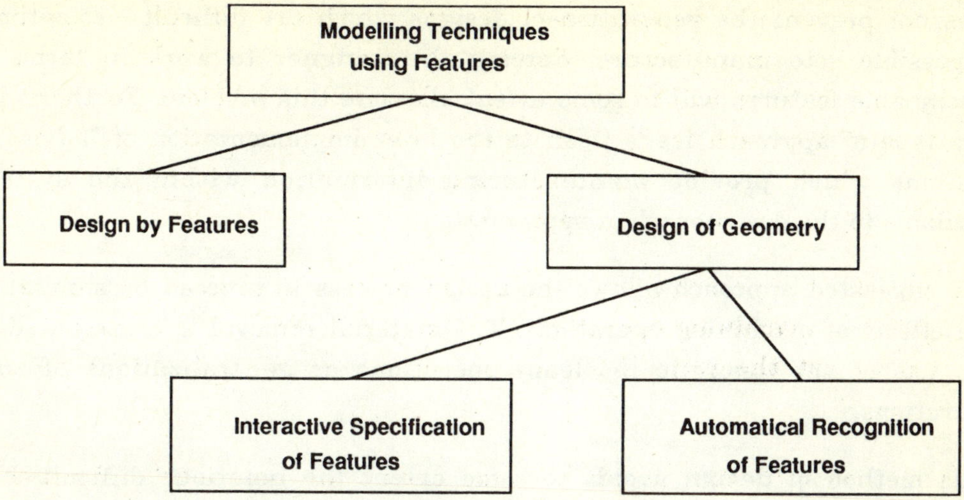

Figure 6-3: Methods for handling features

In the limited time scale of the CAD*I Project, the complexity of automatic methods was avoided by requiring the designer to work in terms of a specific set of features appropriate to the manufacture of parts using conventional machining methods. This leads to a special purpose modelling system. Nevertheless, this approach allows many of the requirements for design by features (in terms of data structures and operations) to be identified.

The requirement for individual organisations to be able to configure the system in terms of classes of features was also neglected to avoid the introduction of further complexity. The proposed simplification allows us to focus on a relatively small number of pre-defined classes of features and hence concentrate on the types of feature operations required for feature-oriented design.

Design for Manufacture of Machined Parts

A good designer will provide a product design which is both functional and feasible to manufacture. This requires some knowledge of manufacturing processes, since even the use of today's sophisticated solid modelling systems does not prevent the generation of designs which are difficult - sometimes impossible - to manufacture. Forcing the designer to work in terms of machinable features will to some extent alleviate this problem. Furthermore, this type of approach leads itself to the later implementation of "advisory" systems which provide manufacturing information within the system, available to the designer when appropriate.

The suggested approach allows the design process to proceed by simulating the effects of machining operations. The material removal is represented by the use of set theoretic (boolean) operations or generalizations of such operations.

This method of design avoids to some extent the notorious difficulties of language and terminology currently experience by designers /PRA-89/.

Before investigating the modelling aspects of machined features some preliminary work is necessary. The required tasks are

I. Identification and classification of the most commonly used machined features in the countries of the European Community.

II. Determination of the most suitable type of solid modeller for implementation of these feature classes.

For example, it is shown in /MWP-84/ that in the United Kingdom engineering industry metal removal operations account for about 70 % of manufacturing operations. Of these, some 60 % are the conventional operations of milling, drilling and turning (see Figure 6-4 and Figure 6-5).

Despite its simplicity, the proposed system is designed to handle interactions between features and patterns of features. A basic demonstration system has been developed. Consideration are given to the problems of interfacing to libraries of standard or user-defined features for design, and to libraries of machine tools and cutting tools for the planning of manufacture.

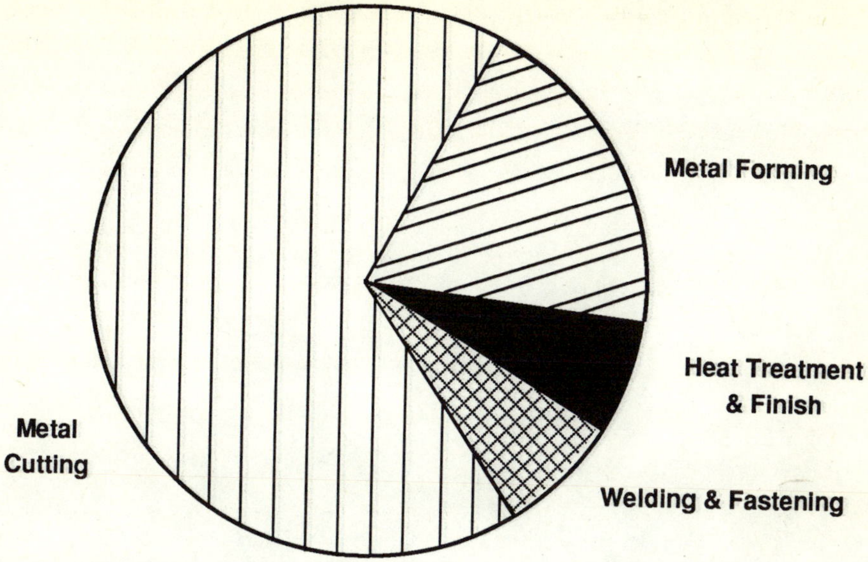

Figure 6-4: Manufacturing Processes in the UK Engineering Industry

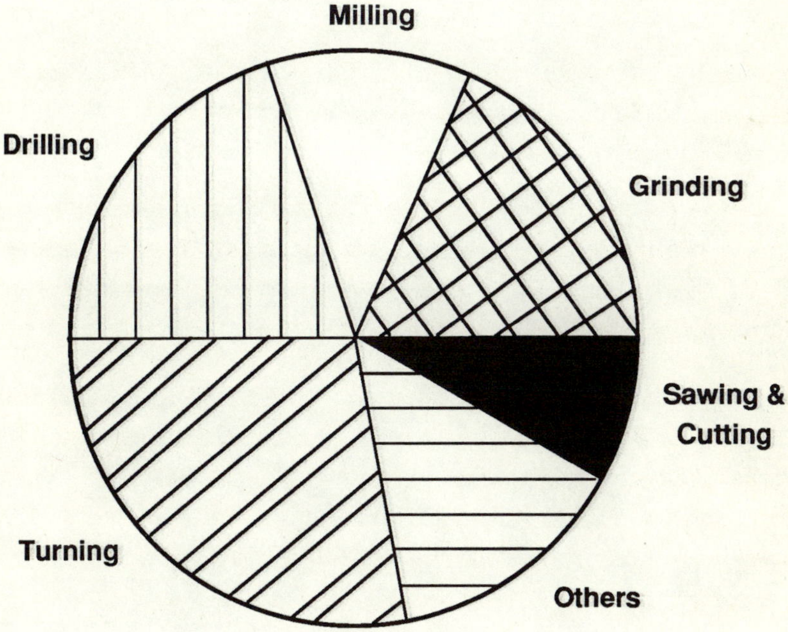

Figure 6-5: Distribution of Metal Removal Operations in Figure 6-4

7. Internal Interfaces

Bernd Pätzold and Stefan Rude

A principle goal of the ESPRIT project CAD*I (CAD Interfaces) is to develop interfaces. The goal of working group 5 "Advanced Modelling" is to develop interfaces for advanced CAD systems. This requires well-defined internal interfaces. Figure 7-1 gives an overview of the relevant interfaces.

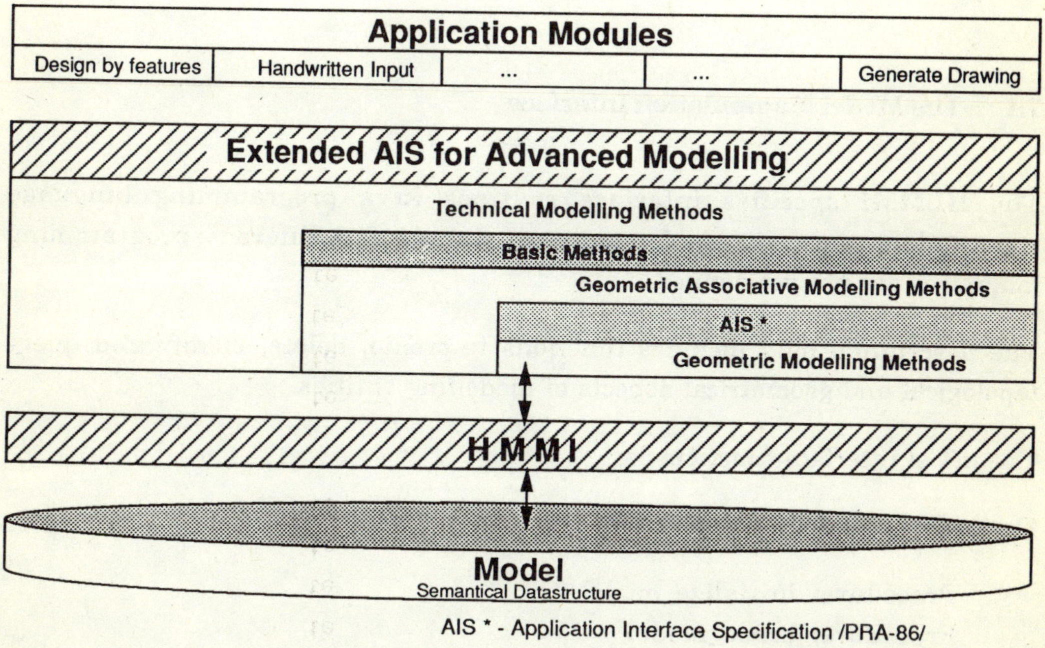

Figure 7-1: CAD Architecture: Internal Interfaces

Section 7.1 describes the syntax of the interface for CAD-data manipulation. This interface is called High Level Model Manipulation Interface (HLMMI). This interface describes an external schema for a solid model based on the B-Rep approach. The purpose of this interface to allow the implementation of modelling operations to remain independent of the underlying solid modelling data structure.

Section 7.2 describes the Geometric-Associative Interface (GAI). Together with the set of AIS-Routines, they form the primitive operations of all other internal interfaces.

Section 7.3 describes the Technical Modelling Operations. They form the Extended AIS for Advanced Modelling together with the Basic Operations.

The DTT-Modeller with the design by features approach and the "Handsketch" Input System (HIS) are two applications which are based on these internal interfaces.

7.1 The Model Manipulation Interface

The HLMMI specifies interface routines in a programming language independent manner. Implementation details for different programming languages are not yet defined.

The description only includes functions to create, delete, modify and query topological and geometrical aspects of modelling entities.

The list of interface functions is:

Group 1: Model Management

Procedure: initialize_model
Procedure: erase_model
Procedure: load_model
Procedure: save_model

Group 2: Create Entity or Associativity

2.1 Create Topology
Procedure: create_body
Procedure: create_shell
Procedure: create_face
Procedure: create_loop
Procedure: create_edge
Procedure: create_vertex

Procedure: create_wire_face
Procedure: create_wire_loop
Procedure: create_wire_edge
Procedure: create_wire_vertex
Procedure: create_wire_face_set
Procedure: create_wire_edge_set

2.2. Create Geometry
Procedure: create_surface
Procedure: create_curve
Procedure: create_vector
Procedure: create_matrix

2.3 Create Associativity
Procedure: create_associativity
Procedure: create_edge_associativity
Procedure: create_wire_edge_associativity

Group 3. Delete Entity or Associativity

Procedure: delete_entity
Procedure: delete_associativity

Group 4: "Get" Functions

4.1 Get Data
Procedure: get_surface_data
Procedure: get_curve_data
Procedure: get_vector_data
Procedure: get_matrix_data
Procedure: get_face_data

4.2. Primitive Queries
Procedure: get_face
Procedure: get_loop
Procedure: get_edge
Procedure: get_vertex
Procedure: get_wire_face
Procedure: get_wire_loop

Procedure: get_wire_edge
Procedure: get_wire_vertex
Procedure: get_wire_face_set
Procedure: get_wire_edge_set

4.3. Adjacency Relationships
Procedure: get_faces_of_face
Procedure: get_loops_of_face
Procedure: get_edges_of_face
Procedure: get_vertices_of_face
Procedure: get_loops_of_loop
Procedure: get_edges_of_loop
Procedure: get_vertices_of_loop
Procedure: get_faces_of_vertex
Procedure: get_loops_of_vertex
Procedure: get_edges_of_vertex
Procedure: get_vertices_of_vertex

Group 5: Modify Data

Procedure: modify_face
Procedure: modify_surface
Procedure: modify_curve
Procedure: modify_vector
Procedure: modify_matrix

7.1.1 Examples of the HLMMI

Model Management:

Procedure:	**initialize_model**
Number:	101
Purpose:	Initialize the data structure.
Parameters:	none
Input:	- none -
Output:	- none -
Description:	The dynamic data structure is initialized. All pointers are set to NIL.
Errors:	none

Procedure: **load_model**

Number: 103

Purpose: Retrieve the model from a file.

Parameters: (name, object,error)

Input: name - of data type "String"
Name of sequential file from which the model is to be read.

Output: object - of data type "Ident"
Name of the object read from the file

error Integer error code

Description: A sequential file is opened for reading. The model data will be read from the file into the dynamic data structure without verifications. The name of the highest hierarchical object of the model is returned.

Errors: 1031 - File does not exist.

"Get" Functions:

Procedure: **get_edge**

Number: 408

Purpose: get the adjacent entities of a edge :
The sysnames of l_loop, r_loop, cwre, ccre, cwle, ccle, p_vertex, n_vertex, curve and part.

Parameters: (sysname,l_loop,r_loop,cwer,ccre,cwle,ccle, p_vertex,n_vertex,curve,part,error)

Input: sysname - Type Ident
The sysname of the edge element.

Output: l_loop - Type Ident
The sysname of the adjacent l_loop entity, zero if no l_loop entity exists.

r_loop - Type Ident
The sysname of the adjacent r_loop entity, zero if no r_loop entity exists.

cwre - Type Ident
The sysname of the adjacent cwre entity, zero if no cwre entity exists.

ccre - Type Ident
The sysname of the adjacent ccre entity, zero if no ccre entity exists.

cwle - Type Ident
The sysname of the adjacent cwle entity, zero if no cwle entity exists.

ccle - Type Ident
The sysname of the adjacent ccle entity, zero if no ccle entity exists.

p_vertex - Type Ident
The sysname of the adjacent p_vertex entity, zero if no p_vertex entity exists.

n_vertex - Type Ident
The sysname of the adjacent n_vertex entity, zero if no n_vertex entity exists.

curve - Type Ident
The sysname of the adjacent curve entity, zero if no curve entity exists.

part - Type Ident
The sysname of the adjacent part entity, zero if no part entity exists.

The different associativities are shown below.

Description: If the edge entity "sysname" exists, the sysnames of the adjacent entities are returned. These are:
The sysname of the l_loop, r_loop, cwre, ccre, cwle, ccle, p_vertex, n_vertex, curve and part entity.

Errors:
4081 - Entity does not exist.
4082 - Entity is not an edge entity.

Procedure: get_vertices_of_face

Number: 419

Purpose: gets the sysnames of the adjacent vertex entities to the given face

Parameters: (sysname,vertex,error)

Input: sysname - Type Ident
The sysname of the face entity.

Output: vertex - Type Ident_list
A clockwise sorted list of sysnames of the adjacent vertex entities, NIL if no adjacent vertex entities exist.

error Integer error code

Description: If the face entity sysname exists, a clockwise sorted list of the adjacent vertex entities is returned.
get_vertices_of_face

Errors:
4191 - Entity does not exist
4192 - Entity is not a face entity

7.1.2 Programming with the HLMMI

The procedures of the HLMMI are used to program basic modelling operations independently of the implementation of the underlying solid modelling data structure. The following short example describes the implementation of a basic Application Interface (AI) function using the HLMMI. The selected AI-procedure make_solid_object is an Euler operation for B-rep solid modelling. The following describes the implementation of this operator in "pseudo-Pascal".

```
PROCEDURE make_solid_object (      user_name: string;
                               VAR object_name,
                                   face_name,
                                   loop_name,
                                   vertex_name: ident;
                                   error: Error_code);

{ Application Interface procedure: make_solid_object (AIS Gr. 7)
  A new object, consisting of a single shell, face and vertex is
  created.The name of the new object and its associated entities
  is returned. }

%INCLUDE ´global_model_declaration´;
VAR  error: error_code;
BEGIN
     create_body(user_name,body_name);
     create_face(undefined,face_name);
     create_associativity(body_name,face_name,error);
     create_loop(loop_name);
     create_associativity(face_name,loop_name),error);
     create_vertex(vertex_name);
     create_associativity(loop_name,vertex_name,error);
END;
```

7.2 Geometric Associative Interface (GAI)

The version presented in this book is an overview of the first revision of the interface specification based on the collaborative work of CIT and UKA. Detailed specifications are available in /GAR-89/.

Group 1 - Test Functions

Geometric Functions

ga_test_parallel

Test the parallelism of two elements, face/face, face/edge or edge/edge. The face must be planar.

ga_test_angular

Test the angle between two elements face/face, face/edge or edge/edge. The face must be planar.

ga_test_coaxial

Test the coaxiality of two elements face/face or face/edge. The faces must be cylindrical or conical and the edges must be straight line segments.

ga_test_matr_edge

Test if two parallel edges contained in parallel faces have the same material side.

ga_test_matr_face

Test if two parallel faces have material in the same direction.

Metric Functions

ga_calculate_distance

Calculate the distance between parallel entities or the shortest distance between them

ga_calculate_distancevector

Calculate the common normal vector of two parallel entities. The vector is oriented towards the action entity (translation vector).

ga_calculate_angle

Calculates the angle between two "angular" entities.

ga_calculate_angleaxis

Calculates the direction vector of an axis of rotation.

ga_calculate_radius

Calculates the radius of a geometric associativity "coaxial".

ga_calculate_radiusaxis

Calculates the direction vector of an axis of symmetry.

Group 2 - Insertion Functions

Geometric Functions

ga_ins_parallel

Creates a geometric associativity "parallel" when the two given entities are parallel.

ga_ins_angular

Creates a geometric associativity "angular" when the two given entities are angular.

ga_ins_coaxial

Creates a geometric associativity "coaxial"
when the two given entities are coaxial.

Metric Functions

ga_ins_distance

Creates a metric constraint in addition to the geometric associativity "parallel".

ga_ins_angle

Creates a metric constraint in addition to the geometric associativity "angular".

ga_ins_radius

Creates a metric constraint in addition to the geometric associativity "coaxial".

ga_ins_shortest_distance

point/face, point/edge, point/point or two oblique- angled edges.

Group 3 - Modification Functions

Geometric Functions

ga_mod_matr_edge

Changes the material side of one of the parallel faces in relation to an edge by rotating one entity.

ga_mod_matr_face

Changes the material direction of one of the parallel faces by rotating one entity.

Metric Functions

ga_mod_distance

Changes the distance parameter of a geometric associativity: Test if the proposed modification is possible. This involves both the topological and geometrical constraints of the entity. If the modification is possible, then the action element and all its related entities are changed by translating it by the given amount.

ga_mod_angle

Changes the distance parameter of a geometric associativity: Test if the proposed modification is possible. This involves both the topological and geometrical constraints of the entity. If the modification is possible, then the action element and all its related entities are changed by rotating it by the given amount.

ga_mod_radius

Changes the distance parameter of a geometric associativity: Test if the proposed modification is possible. This involves both the topological and geometrical constraints of the entity. If the modification is possible, then the action element and all its related entities are changed by dimensioning it with the given amount.

ga_mod_shortest_distance

Changes the distance parameter of a geometric associativity: Test if the proposed modification is possible. This involves both the topological and geometrical constraints of the entity. If the modification is possible, then the action element and all its related entities are changed by translating it by the given amount.

Group 4 - Deletion Functions

ga_del

Deletes both a metric constraint and its associativity or just the specified associativity.

ga_del_constraint

Only deletes the specified metric constraint. The geometric associativity is not affected.

7.3 Technical Modelling Interface

The following is an overview of the draft version of the technical modelling interface. A detailed specification is available in /GAR-89/.

Group 1 - Test Functions

ta_test_taentity

Test if an entity has a technical associativity(s) belonging to it

ta_test_auxentity

Test if an entity has technical auxiliary entities such as construction lines or spaces belonging to it

ta_test_axis

Test if the entity has an axis belonging to it

Group 2 - Insertion Functions

2.1 Technical auxiliary entity insertion

ta_ins_conspace

creates a volume as a design space

ta_ins_conface

creates a face as a construction plane

ta_ins_conline

creates a line as a construction line

ta_ins_conpoint

creates a construction point

ta_ins_axis

creates an axis

2.2 Create Technical associativities

ta_ins_prod

Create a "product" entity. Attach assemblies or parts to it

ta_ins_assembly

Create an "assembly" entity. Attach parts to it

ta_ins_part

Create a "part" entity. Attach faces or form features to it

ta_ins_formfeature

Create a "form feature" entity. Attach faces or edges to it

Group 3 - Modification Functions

3.1 Geometric transformations position modification

ta_mod_trans

translate all assembly entities except for the reference entity

ta_mod_rot

rotate all assembly entities except for the reference entity

3.2 "Technical" modification of position and orientation

ta_mod_flush

place an entity flush to another

ta_mod_alignface

aligns two entities on a given surface

ta_mod_alignedge

aligns two entities on a given edge

ta_mod_coaxial

makes two cylindrical or conical entities coaxial
and inserts the axis as an explicit entity

ta_mod_adjacent

places one part on top of another after identification of the
two contacting surfaces

Group 4 - Deletion Functions

ta_del_taentity

a technical associativity entity such as "flush", "coaxial", "product" or
"assembly"
is deleted

ta_del_auxentity

an auxiliary technical entity such as "construction line" or
"construction point" is deleted

8. References

/ALD-88/ Aldefeld, B.: Variation of Geometrics Based on a Geometric-Reasoning Method, CAD, Vol. 20, No. 3, Butterworth, Guildford/England, April 1988

/AND-85/ Anderl, R.: Fertigungsplanung durch die Simulation von Arbeitsvorgängen auf der Basis von 3D-Produktmodellen, VDI-Fortschrittsberichte Reihe 10, No. 40, VDI-Verlag, Düsseldorf/Germany, 1985

/AND-89/ Anderson, J.: ICAD - "A Knowledge-Based ICAD System", Design Automation Workshop, CAM-I Report P-89-GM-01, Arlington, Texas/U.S.A., 1989

/ARB-87/ Arbab, F.: A Paradigm for Intelligent CAD. In: P.J.W. ten Hagen, T. Tomiyama (Eds.) : Intelligent CADD Systems I. Theoretical and Methodological Aspects. Springer-Verlag, Berlin/Germany, 1987, pp. 20 - 37

/BJO-78/ Bjørke, O.: Advanced Production System - CAD/CAM for the Future, pp. 64 - 81, pp. 261-267, Tapir Pub./Norway, 1987

/BEN-90/ Benz, T.: Funktionsmodelle in CAD-Systemen, VDI-Verlag, Düsseldorf/Germany, 1990

/BRE-88/ Breitling, F.: Wissensbasiertes Konstruktionssystem, Zeitschrift für wirtschaftliche Fertigung und Automatisierung (ZwF) 83 (1988) 11

/CAD-89/ CAD*I Status Report No. 5, ESPRIT Project CIM 322: CAD Interfaces. KfK-PFT 145, Kernforschungszentrum Karlsruhe/Germany, 1989

/CAM-88/ Shah, J.J., Sreevalsan, P., Rogers, M., Billo, R., Mathew, A.: Current Status of Features Technology: Revised Report, CAM-I Report R-88-GM-04.01, Arlington, Texas/USA., 1988

/CDG-84/ Cugini, M., Devoti, C., Galli, P.: Language for Definition and Manipulation of Parametric Graphic Representations, Proc. CAD 84, Brighton/England, 1984

/CDG-85/ Cugini, M., Devoti, C., Galli, P.: System for Parametric Definition of Engineering Drawings, Proc. MICAD 85, 1985

/COH-88/ Conrads, G., Hornung, V.: Neue Möglichkeiten durch objektorientierte Arbeitstechniken bei der Arbeit mit CAD-Systemen, VDI-Berichte Nr.700, VDI-Verlag, Düsseldorf/Germany 1989

/FER-89/ Feranti (ed.): Pro/ENGINEER Concepts, Feranti International GmbH, Frankfurt/Germany, 1989

/FIT-81/ Fitzgerald, W.J.: Using Axial Dimensions to Determine the Proportions of Line Drawings in Computer Graphics, CAD Vol.13, No.6, Butterworth, Guildford/England, Nov. 1981

/GAR-86/ Grabowski, H., Anderl, R., Rude S.: Technical Term Dictionary with Related Technical and Geometrical Semanties, Project Report, ESPRIT Project 322, CAD-Interfaces (CAD*I), KfK-PFT 133, Kernforschungszentrum Karlsruhe/Germany, 1989

/GAR-89/ Grabowski, H., Anderl, R., Rude, S.: Design by Technical Terms, Project Report, Esprit Project 322, CAD-Interfaces (CAD*I), CAD*I.WG5.UKA.0003.88, Karlsruhe/Germany, 1989

/GRB-89/ Grabowski, H., Benz, T.: Modelling the Design Methodology, IFIP WG 5.2, Preprints of 3rd Working Conference, Osaka/Japan, 1989

/GRR-88a/ Grabowski, H., Rude, S.: Methoden zur Lösungsfindung in CAD-Systemen, International Conference on Engineering Design (ICED), Budapest/Hungary, 1988

/GRR-88b/ Grabowski, H., Rude, S.: Intelligent CAD-Systems Based on Technical Associative Modelling. In: W. Straßer, H.-P. Seidel (Eds.): Theory and Practice of Geometric Modeling, Springer-Verlag,Berlin/Germany, 1989, pp. 451-467

/GRR-90/ Grabowski, H., Rude, S.: Methodisches Entwerfen auf der Basis zukünftiger CAD-Systeme, VDI-Berichte Nr. 812, VDI-Verlag, Düsseldorf/Germany, 1990

/GRS-84/ Grabowski, H., Seiler, W.: Techniques, Operations and Models for Functional and Preliminary Design Phases, Proc. ISDS Tokyo/Japan, 1984

/GZS-88/ Gossard, D.C., Zuffante, R.P., Sakurai, H.: Representing Dimensions, Tolerances and Features in MCAE Systems, IEEE CG & A, Vol. 8 No. 2, March 1988

/HEI-90/ Heidrich, R.: Ein Beitrag zur Konzeption und Anwendung parametrisierter, integrierter Produktmodelle in CAD-Systemen, VDI-Berichte, Reihe 20, Nr. 27, VDI-Verlag, Düsseldorf/Germany, 1990

/HIB-78/ Hillyard, R.C., Braid, I.C.: Analysis of Dimensions and Tolerances in Computer-Aided Mechanical Design, CAD Vol. 10, No. 3, Butterworth, Guildford/England, May 1978

/HOF-87/ Hoffmann, H.: Methoden zur anwendungsorientierten Gestaltung von Geometrieverarbeitungssystemen, Produktionstechnik-Berlin, Nr. 61, Hanser-Verlag, München/Germany, 1987

/HOK-77/ Hosaka, M., Kimura, F.: An Interactive Geometrical Design System with Handwritten Input, IFIP Congress Proc. 1977, pp. 167-172

/ICA-81/ Integrated Computer Aided Manufacturing (ICAM): Function Modeling Manual (IDEF0), Materials Lab, AF Wright, Aeronautical Laboratories/U.S.A., 1981

/IMP-89/ IMPPACT Project Consortium: Integrated Modelling of Products and Processes using Advanced Computer Technologies (IMPPACT), ESPRIT Project No. 2165, ECW Brussels/Belgium, 1989

/ISO-89/ Schenk, D.A.: Information Modeling Language EXPRESS, ISO TC184/SC4/WG1, N362, 1989

/JNR-85/ Jansen, H., Nullmeier, E., Roediger, K.-H., Handsketching as a Human Faction Aspect in Graphical Interaction, Computer & Graphics Vol. 9, No. 3, pp. 195-210, 1985

/KAN-88/ Kandziora, B.: CAD/CAM-System zur Planung und Simulation automatisierter Montagevorgänge, VDI-Fortschrittsberichte, Reihe 20, Nr. 9, VDI-Verlag, Düsseldorf/Germany, 1988

/KDY-85/ Kurz, O., Daßler, R., Yaramanoglu, N.: Variable Geometriemodelle als Basis für eine rationelle rechnerunterstützte Konstruktionsmethode, VDI-Berichte Nr. 570.5, VDI-Verlag, Düsseldorf/Germany, 1985

/KJT-88/ Krause, F.-L., Jansen, H., Timmermann, M.: Handskizzenentwurf - eine neue Möglichkeit zur Gestaltung der Benutzeroberfläche von CAD-Systemen, GI-Fachgespräch FG 4.2.1., Berlin/Germany, 1989

/LIG-82/ Light, R., Gossard, D.: Modification of Geometric Models Through Variation Geometry, CAD Vol. 14, No. 4, Butterworth, Guildford/England, July 1982

/LIN-86/ Lindem, T.: Integrated Design and Manufacturing, VDI Bericht 611, VDI-Verlag, Düsseldorf/Germany, 1986

/MAC-88/ Macilwain, C.: CAD's Fourth Generation, The Engineer, October 1988

/MWP-84/ Fifth Survey of Machine Tools and Production Equipment, Metal Working Production, London/England, 1984

/NAR-68/ Narasiham, R.: Syntax Directed Interpretation of Classes of Pictures, Comm. ACM 9, pp. 166-179, 1989

/PAB-86/ Pahl, G., Beitz, W.: Konstruktionslehre, Springer-Verlag, Berlin/Germany, 1986

/PRA-87/ Pratt, M.J.: Form Features and their Applications in Solid Modelling, Course Notes Vol. 26, Advanced Topics in Solid Modelling, Siggraph 87 Conf., Anaheim, California/U.S.A., 1987

/PRA-88/ Pratt, M.J.: The Form Feature Concept for Advanced Modelling, 3th Int. Workshop on CAD-Interfaces, Lyngby/Denmark, 1988

/PRA-89/ Pratt, M.J.: Conceptual Design of a Feature Oriented Solid Modeller, Internal Report , General Electric Corporate Research & Development Center, Schenectady, New York/USA., 1989

/ROS-85/ Ross, D.T.: Applications and Extensions of SADT; Computer, No. 4, 1985

/RSV-89/ Roller, D., Schonek, F., Verroust, A.: Dimension-Driven Geometry in CAD: A Survey. In: W. Straßer, H.-P. Seidel (Eds.): Theory and Practice of Geometric Modeling, Springer-Verlag, Berlin/Germany, 1989, pp. 509-523

/RUT-85/ Rutz, A.: Konstruieren als gedanklicher Prozeß, Dissertation, TU München/Germany, 1985

/SCH-89/ Schlechtendahl, E.G.(Ed.): CAD Data Transfer for Solid Models, Res. Rep. Esprit Project 322 (CAD*I) Vol. 3, Springer-Verlag, Berlin/Germany, 1989

/SEI-85/ Seiler, W.: Technische Modellierungs- und Kommunikationsverfahren für das Konzipieren und Gestalten auf der Basis der Modellintegration, VDI-Fortschrittsberichte, Reihe 10: Angewandte Informatik, Nr. 49, VDI-Verlag, Düsseldorf/Germany, 1985

/SKJ-87/ Spur, G., Krause, F.-L., Jansen, H.: Verfahren zur automatischen Zeichnungsinterpretation für CAD-Prozesse, Zeitschrift für wirtschaftliche Fertigung und Automatisierung (ZwF) 82 (1987) 5

/SUN-86/ Sunde, G.: Specification of Shape and Dimensions and other Geometry Constraints, IFIP WG5.2 on Geometric Modeling, Rensselaerville New York/USA, 1988

/SUN-87/ Sunde, G.: A CAD system with Declarative Specification of Shape. In: P. J. W. ten Hagen, T, Tomiyama (Eds.): Intelligent CAD Systems I. Theoretical and Methodological Asoects. Springer-Verlag, Berlin/Germany, 1987, pp. 90-104

/TUW-88/ Turner, J.U., Wozny, M.: A Mathematical Theory of Tolerances, IFIP Geometric Modelling for CAD Applications, North-Holland, Amsterdam/The Netherlands, 1988

/VDI-85/ VDI (ed.): Methodik zum Entwickeln und Konstruieren technischer Systeme und Produkte, VDI-Richtlinie 2221, Entwurf VDI-Handbuch Konstruktion, Beuth-Verlag, Berlin/Germany, 1985

/WIJ-87/ Wijk, J.: A Solid Modelling Language, TNO-IW, Delft/The Netherlands, 1987